Springer Theses

Recognizing Outstanding Ph.D. Research

Aims and Scope

The series "Springer Theses" brings together a selection of the very best Ph.D. theses from around the world and across the physical sciences. Nominated and endorsed by two recognized specialists, each published volume has been selected for its scientific excellence and the high impact of its contents for the pertinent field of research. For greater accessibility to non-specialists, the published versions include an extended introduction, as well as a foreword by the student's supervisor explaining the special relevance of the work for the field. As a whole, the series will provide a valuable resource both for newcomers to the research fields described, and for other scientists seeking detailed background information on special questions. Finally, it provides an accredited documentation of the valuable contributions made by today's younger generation of scientists.

Theses may be nominated for publication in this series by heads of department at internationally leading universities or institutes and should fulfill all of the following criteria

- They must be written in good English.
- The topic should fall within the confines of Chemistry, Physics, Earth Sciences, Engineering and related interdisciplinary fields such as Materials, Nanoscience, Chemical Engineering, Complex Systems and Biophysics.
- The work reported in the thesis must represent a significant scientific advance.
- If the thesis includes previously published material, permission to reproduce this must be gained from the respective copyright holder (a maximum 30% of the thesis should be a verbatim reproduction from the author's previous publications).
- They must have been examined and passed during the 12 months prior to nomination.
- Each thesis should include a foreword by the supervisor outlining the significance of its content.
- The theses should have a clearly defined structure including an introduction accessible to new PhD students and scientists not expert in the relevant field.

Indexed by zbMATH.

Katsuki Kinjo

Nuclear Magnetic Resonance Study on Multiple Superconducting Phases in UTe_2

Doctoral Thesis accepted by
Kyoto University, Kyoto, Japan

Springer

Author
Dr. Katsuki Kinjo
Tohoku University
Sendai, Japan

Supervisor
Prof. Kenji Ishida
Department of Physics
Graduate School of Science
Kyoto University
Kyoto, Japan

ISSN 2190-5053 ISSN 2190-5061 (electronic)
Springer Theses
ISBN 978-981-97-8645-9 ISBN 978-981-97-8646-6 (eBook)
https://doi.org/10.1007/978-981-97-8646-6

This Springer imprint is published by the registered company Springer Nature Singapore Pte Ltd.
The registered company address is: 152 Beach Road, #21-01/04 Gateway East, Singapore 189721, Singapore

Supervisor's Foreword

This Ph.D. thesis describes an investigation conducted by Dr. Katsuki Kinjo during his doctoral program at the Graduate School of Science, Kyoto University. In his graduate study, he achieved significant contributions to the research of a new uranium-based heavy-fermion superconductor UTe_2, which was discovered by Dr. Shen Ran et al. in University of Maryland at the end of 2018. After this, UTe_2 has attracted much attention since it has been regarded as a promising candidate of spin-triplet superconductivity from the robustness of superconductivity in the magnetic field greater than 16 T and the multiple superconducting (SC) phases under hydrostatic pressure. Dr. Kinjo made several important discoveries on the magnetic properties on UTe_2, particularly in the SC state. As the supervisor of the doctoral course, here I introduce his remarkable discoveries on $UTe_{2.}$.

The first one is about the high magnetic field SC state. Although this state was confirmed by the resistivity measurements, it was not thoroughly investigated before 2023, whether this state is bulk properties or not. Dr. Kinjo performed AC magnetic susceptibility and nuclear magnetic resonance measurements up to 24.8 T and found that the high-field SC state has bulk nature, and that its SC character, particularly spin state of the SC pairing, is different from that in the low-field SC state. The high-field SC phase diagram is quite similar to that reported by A. Rosuel et al., determined with the specific heat measurements.

The other is about the magnetic and SC properties under pressure. Dr. Kinjo performed NMR measurements in the normal and SC states under pressure. He found that superconductivity in UTe_2 is observed only in the heavy-fermion state and that Fermi-surface character is suddenly changed at the critical pressure, above which superconductivity was not observed. In addition, he found that spin susceptibility when $H \parallel b$ is unchanged in the high-temperature SC state under pressure, which is the different spin state than that in the SC state at ambient pressure. Furthermore, he discovered that the spin susceptibility decreases with spin modulation below the SC-SC transition under pressure and suggested the spin-triplet SC state with SC spin parallel to b axis and the remaining spin degrees of freedom in the SC state under pressure.

On the contrary to the conventional spin-singlet superconductors, the spin-triplet superconductors have spin degrees of freedom, which are believed to play an important role in the realization of SC multiphases. As mentioned above, Dr. Kinjo gave definitive information about the spin state in several SC phases. I hope that a result of Dr. Kinjo's efforts gives a valuable clue to unveil the mystery of the SC state in $UTe_{2.}$.

Kyoto, Japan
November 2023

Prof. Kenji Ishida

Parts of this thesis have been published in the following journal articles

1. K. Kinjo et al., "Drastic change in magnetic anisotropy of under pressure revealed by ^{125}Te-NMR," Phys. Rev. B **105**, L140502 (2022).
2. K. Kinjo et al., "Change of superconducting character in UTe_2 induced by magnetic field," Phys. Rev. B **107**, L060502 (2022).
3. K. Kinjo et al., "Superconducting spin reorientation in spin-triplet multiple superconducting phases of UTe_2," Sci. Adv. **9**, eadg2736 (2023).

Acknowledgements

I would like to express sincere gratitude to Prof. Kenji Ishida, my supervisor, and S. Kitagawa for their helpful suggestions and fruitful discussions throughout my 6 years in this laboratory. I greatly acknowledge Mr. H. Fujibayashi and H. Matsumura, as my great colleagues in research life, NMR measurements, and UTe_2. This research on UTe_2 could not be done without a high-quality sample. I would like to express my gratitude to the sample provider, Prof. D. Aoki at Tohoku University. I would like to thank Dr. Kimata at IMR for his help in the measurement of the high-magnetic field in Chap. 6. Our conclusions are strongly supported by theorists Prof. Y. Yanase, Prof. K. Machida, and Prof. S. Fujimoto for their strong theoretical implications. I would like to give a great acknowledgment for indispensable discussions with them. I would like to thank other scientists for studying UTe_2 with us, especially A. Miyake, J. P. Brison, and J. Floquet.

For my research life in QM Lab., I acknowledge Prof. Y. Maeno and Prof. S. Yonezawa for variable discussions. I am also supported by A. Ikeda, for his keen insight. I appreciate all the other members and the former members of the QM Lab. for valuable discussions, useful support in many aspects, and sharing a good time.

Finally, I sincerely appreciate my family, who always support and encourage me during my life.

Contents

Chapter 1
Introduction

1.1 Spin-Triplet Superconductivity

1.1.1 Superconductivity

Superconductivity is the phenomenon of certain materials exhibiting zero resistivity and the expulsion of magnetic fields below a characteristic temperature (T_c). The superconductivity phenomenon was discovered in 1911 by Dutch physicist Heike Kamerlingh Onnes. The microscopic theory for superconductivity was proposed in 1957 by John Bardeen, Leon N. Cooper, and Robert Schrieffer. This BCS theory explained superconductivity as a superfluid of Cooper pairs, pairs of electrons interacting through the exchange of phonons. In BCS theory, two electrons are paired; if SOC is weak and parity is conserved, the wave function can be written with the spin and orbital parts separated as follows

$$\Psi\left(\boldsymbol{r}_1, \sigma_1; \boldsymbol{r}_2, \sigma_2\right) = \varphi\left(\boldsymbol{r}_1 - \boldsymbol{r}_2\right) \chi\left(\sigma_1, \sigma_2\right) \tag{1.1}$$

Since the two-body wave function Ψ is antisymmetric with respect to electron replacement, the spin part of the Cooper pair $\chi\left(\sigma_1, \sigma_2\right)$ becomes a spin singlet state with spin $S = 0$ when the orbital part is an even function,

$$|S = 0, S_z = 0\rangle = \frac{1}{\sqrt{2}}(|\uparrow\downarrow\rangle - |\downarrow\uparrow\rangle). \tag{1.2}$$

K. Kinjo, *Nuclear Magnetic Resonance Study on Multiple Superconducting Phases in UTe2*, Springer Theses, https://doi.org/10.1007/978-981-97-8646-6_1

On the other hand, when the orbital part is an odd function, $\chi(\sigma_1, \sigma_2)$ becomes a spin-triplet state with $S = 1$,

$$|S = 1, S_z = 1\rangle = |\uparrow\uparrow\rangle \tag{1.3}$$

$$|S = 1, S_z = 0\rangle = \frac{1}{\sqrt{2}}(|\uparrow\downarrow\rangle + |\downarrow\uparrow\rangle) \tag{1.4}$$

$$|S = 1, S_z = -1\rangle = |\downarrow\downarrow\rangle. \tag{1.5}$$

1.1.2 Order Parameter of Superconductivity

Following the BCS theory, we derive the superconducting order parameter. In general, the Hamiltonian of the interacting electron system is as follows,

$$\mathcal{H} = \sum_{k}\sum_{\sigma} \xi_k c^{\dagger}_{k,\sigma} c_{k,\sigma} + \frac{1}{2}\sum_{k,k'}\sum_{\sigma,\sigma'} V\left(k, k'\right) c^{\dagger}_{k,\sigma} c^{\dagger}_{-k,\sigma'} c_{-k',\sigma'} c_{k',\sigma}. \tag{1.6}$$

The first term is the kinetic energy of electrons and the second term is electron-electron interaction. In this case, we are considering a spin-independent interaction. The BCS theory considers pairs of electrons, i.e., $\langle c^{\dagger} c^{\dagger}\rangle$ becomes non-zero, and the system spontaneously breaks the U(1) gauge symmetry. Then, mean-field Hamiltonian is given by

$$\mathcal{H} = \sum_{\boldsymbol{k}}\sum_{\sigma} \xi_{\boldsymbol{k}} c^{\dagger}_{\boldsymbol{k},\sigma} c_{\boldsymbol{k},\sigma} + \frac{1}{2}\sum_{\boldsymbol{k}}\sum_{\sigma,\sigma'} \left[\Delta_{\sigma,\sigma'}(\boldsymbol{k}) c^{\dagger}_{\boldsymbol{k},\sigma} c^{\dagger}_{-\boldsymbol{k},\sigma'} + \text{h.c.}\right]$$
$$-\frac{1}{2}\sum_{\boldsymbol{k}}\sum_{\sigma,\sigma'} \Delta_{\sigma,\sigma'}(\boldsymbol{k}) \left\langle c^{\dagger}_{\boldsymbol{k},\sigma} c^{\dagger}_{-\boldsymbol{k},\sigma'}\right\rangle, \tag{1.7}$$

where $\Delta_{\sigma,\sigma'}(\boldsymbol{k})$ is the energy gap for quasi-particle and order parameter for superconductivity written as

$$\Delta_{\sigma,\sigma'}(\boldsymbol{k}) = \sum_{\boldsymbol{k}'} V\left(\boldsymbol{k}, \boldsymbol{k}'\right) \langle c_{-\boldsymbol{k}',\sigma'} c_{\boldsymbol{k}',\sigma}\rangle \tag{1.8}$$

Next, we consider the spin-dependent interaction. When the system is symmetric in the spin space, V is written as

$$V_{\alpha,\beta,\gamma,\delta}\left(\boldsymbol{k}, \boldsymbol{k}'\right) = V_1\left(\boldsymbol{k}, \boldsymbol{k}'\right) \delta_{\alpha\delta}\delta_{\beta\gamma} + V_2\left(\boldsymbol{k}, \boldsymbol{k}'\right) (\boldsymbol{\sigma})_{\alpha\delta} \cdot (\boldsymbol{\sigma})_{\beta\gamma}, \tag{1.9}$$

where $\boldsymbol{\sigma} = \left(\hat{\sigma}_x, \hat{\sigma}_y, \hat{\sigma}_z\right)$ is the Pauli matrix and written as

$$\hat{\sigma}_x = \begin{pmatrix} 0 & 1 \\ 1 & 0 \end{pmatrix}, \quad \hat{\sigma}_y = \begin{pmatrix} 0 & -i \\ i & 0 \end{pmatrix}, \quad \hat{\sigma}_z = \begin{pmatrix} 1 & 0 \\ 0 & -1 \end{pmatrix} \tag{1.10}$$

Using a matrix, we get

$$H = \frac{1}{2} \sum_k \left(c^\dagger_{\boldsymbol{k},\uparrow} \; c^\dagger_{\boldsymbol{k},\downarrow} \; c_{-\boldsymbol{k},\uparrow} \; c_{-\boldsymbol{k},\downarrow} \right) \begin{pmatrix} \xi_k \hat{\sigma}_0 & \hat{\Delta}(\boldsymbol{k}) \\ \hat{\Delta}^*(\boldsymbol{k}) & -\xi_k \hat{\sigma}_0 \end{pmatrix} \begin{pmatrix} c_{\boldsymbol{k},\uparrow} \\ c_{\boldsymbol{k},\downarrow} \\ c^\dagger_{-\boldsymbol{k},\uparrow} \\ c^\dagger_{-\boldsymbol{k},\downarrow} \end{pmatrix} \tag{1.11}$$

Then, gap function $\hat{\Delta}(\boldsymbol{k})$ become 2×2 matrix

$$\hat{\Delta}(\boldsymbol{k}) = \begin{pmatrix} \Delta_{\uparrow\uparrow}(\boldsymbol{k}) & \Delta_{\uparrow\downarrow}(\boldsymbol{k}) \\ \Delta_{\downarrow\uparrow}(\boldsymbol{k}) & \Delta_{\downarrow\downarrow}(\boldsymbol{k}) \end{pmatrix}. \tag{1.12}$$

Spin-Singlet Case

For spin-singlet superconductor, the gap function is scalar then,

$$\hat{\Delta}(\boldsymbol{k}) = \begin{pmatrix} 0 & \psi(\boldsymbol{k}) \\ -\psi(\boldsymbol{k}) & 0 \end{pmatrix} = i\hat{\sigma}_y \psi(\boldsymbol{k}). \tag{1.13}$$

Since spin-part is antisymmetric, $\psi(\boldsymbol{k}) = \psi(-\boldsymbol{k})$ is realized.

Spin-Triplet Case

Spin-triplet Cooper pairs have spin 1, and Eqs. (1.3)–(1.5) are the basis function in the spin state. Traditionally, it is transformed into a convenient basis for rotation in the spin space,

$$\hat{x} = |S_x = 0\rangle = \frac{1}{\sqrt{2}}(-|\uparrow\uparrow\rangle + |\downarrow\downarrow\rangle) \tag{1.14}$$

$$\hat{y} = \left|S_y = 0\right\rangle = \frac{i}{\sqrt{2}}(|\uparrow\uparrow\rangle + |\downarrow\downarrow\rangle) \tag{1.15}$$

$$\hat{z} = |S_z = 0\rangle = \frac{1}{\sqrt{2}}(|\uparrow\downarrow\rangle + |\downarrow\uparrow\rangle) \tag{1.16}$$

Then, order parameter Δs are written in $\boldsymbol{d}$ as follows,

$$\Delta_{\uparrow\uparrow}(\boldsymbol{k})|\uparrow\uparrow\rangle + \Delta_{\downarrow\downarrow}(\boldsymbol{k})|\downarrow\downarrow\rangle + \Delta_{\uparrow\downarrow}(\boldsymbol{k})(|\uparrow\downarrow\rangle + |\downarrow\uparrow\rangle) = \sqrt{2}\left[d_x(\boldsymbol{k})\hat{x} + d_y(\boldsymbol{k})\hat{y} + d_z(\boldsymbol{k})\hat{z}\right]. \tag{1.17}$$

This $\boldsymbol{d}$ is called ***d*-vector**, and d-vector is different notations for the spin-triplet order parameter. Using d-vector, gap functions become

$$\hat{\Delta}(\boldsymbol{k}) = \begin{pmatrix} \Delta_{\uparrow\uparrow}(\boldsymbol{k}) & \Delta_{\uparrow\downarrow}(\boldsymbol{k}) \\ \Delta_{\downarrow\uparrow}(\boldsymbol{k}) & \Delta_{\downarrow\downarrow}(\boldsymbol{k}) \end{pmatrix} = \begin{pmatrix} -d_x(\boldsymbol{k}) + id_y(\boldsymbol{k}) & d_z(\boldsymbol{k}) \\ d_z(\boldsymbol{k}) & d_x(\boldsymbol{k}) + id_y(\boldsymbol{k}) \end{pmatrix} = i(\boldsymbol{d}(\boldsymbol{k}) \cdot \boldsymbol{\sigma})\hat{\sigma}_y. \tag{1.18}$$

1.1.3 Classification of Superconducting State

Considering the Landau theory for phase transition, the ground state has a subgroup of the Hamiltonian of the system. This means that symmetries other than the $U(1)$ gauge can also be broken in superconductors. The superconductivity breaks only $U(1)$ gauge symmetry is called conventional superconductivity, and the other is called **unconventional superconductivity**.

Isortropic Case

First, we consider the isotropic Hamiltonian. For isotropic case, $V(\boldsymbol{k}, \boldsymbol{k}')$ can be decomposed using the spherical harmonic function $Y_{lm}(\hat{\boldsymbol{k}})$,

$$V_{1,2}\left(\boldsymbol{k}, \boldsymbol{k}'\right) = -4\pi \sum_l V_{1,2}^{(l)} \sum_m Y_{lm}(\hat{\boldsymbol{k}}) Y_{lm}^*(\hat{\boldsymbol{k}}'). \tag{1.19}$$

Substituting this into the gap equation, we get an independent gap equation for each (l, m), and

$$\Delta(\boldsymbol{k}) = \Delta^{(l,m)} Y_{lm}(\hat{\boldsymbol{k}}) \tag{1.20}$$

is the solution. In this case, the one with the highest T_c will be realized. Each superconductivity is called s-wave ($l = 0$), p-wave ($l = 1$), d-wave ($l = 2$), f-wave ($l = 3$), and so on. All superconductivity without s-wave breaks the rotational symmetry. Then only s-wave is the conventional superconductivity and the others are the unconventional superconductivity.

Anisotropic Case

To understand the symmetry of the Hamiltonian, the representation theory of groups is a powerful tool. The group of symmetry operations that can be performed on superconducting order parameters $\mathcal{G}$ can be written as

$$\mathcal{G} = SO(3) \otimes \mathcal{P} \otimes SU(2) \otimes \mathcal{T} \otimes U(1) \tag{1.21}$$

by using group of rotation operations $SO(3)$, a spatial inversion $\mathcal{P}$, rotation operations in spin space $SU(2)$, a time reversal operation $\mathcal{T}$ and gauge transformation operations $U(1)$. In the crystal, there is a crystal field that depends on the symmetry of the crystal

Table 1.1 Classification of SC order parameter for point groups with D_{2h}

Even parity			Odd parity		
IR	Basis function	Node	IR	Basis function	node
A_g	$1, k_a^2, k_b^2, k_c^2$	–	A_u	$k_a\hat{a}, k_b\hat{b}, k_c\hat{c}$	–
B_{1g}	$k_a k_b$	Line	B_{1u}	$k_b\hat{a}, k_a\hat{b}$	point $\parallel c$
B_{2g}	$k_a k_c$	Line	B_{2u}	$k_a\hat{c}, k_c\hat{a}$	point $\parallel b$
B_{3g}	$k_b k_c$	Line	B_{3u}	$k_c\hat{b}, k_b\hat{c}$	point $\parallel a$

structure, and therefore the rotational symmetry is broken and the group is rewritten by using a point group G of the crystal as follow,

$$\mathcal{G} = G \otimes SU(2) \otimes \mathcal{T} \otimes U(1). \tag{1.22}$$

In the case of the second-order phase transition, the symmetry of the SC order parameter belongs to one of the irreducible representations (IRs) in the group G. The crystal structure of UTe_2 has the point group D_{2h}, then the allowed order parameters are listed in Table 1.1.

Note that we do not consider the multi-band feature in this section.

1.2 Heavy Fermion System and Kondo State

In this chapter, I briefly introduce the concept of the heavy-fermion system as the Kondo lattice based on Ref. [1].

1.2.1 Local or Itinerant

Heavy electron materials are a class of strongly correlated electron material containing localized magnetic moments which, by entangling with the surrounding electrons, profoundly transform the metallic properties. A heavy fermion metal can develop electron masses 1000 times larger than that of copper, it can also develop unconventional superconductivity, transforms into new states of quantum order, and exhibits quantum critical and topological behavior. In the periodic table, the most strongly interacting electrons reside in orbitals that are well-localized. In order of increasing localization, partially filled orbitals are ordered as follows:

$$5d < 4d < 3d < 5f < 4f \tag{1.23}$$

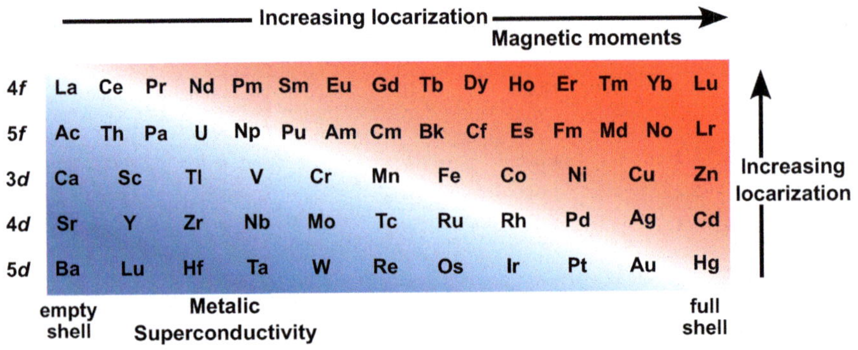

Fig. 1.1 The Kmetko-Smith diagram [2]. This diagram shows a trend toward increased electron localization in d- and f-electron compounds

In addition, when moving along a row of the periodic table, the increasing nuclear charge pulls the orbitals toward the nucleus. These trends are summarized in the "Kmetko-Smith diagram" in Fig. 1.1. The d-orbital metals at the bottom left of this diagram are highly itinerant and exhibit conventional superconductivity. In contrast, in rare earth and actinide metals towards the top right-hand corner, the f-shell electrons are localized, showing magnetism. It is the materials that lie in the cross-over between these two regions that are particularly interesting. For the heavy fermion system, this localized–itinerant duality is the key factor in realizing fascinating phenomena such as superconductivity and magnetism.

1.2.2 Kondo Effect: Localized Moment in the Metal

In rare earth and actinide ions, the orbital and spin angular momentum combine into a single quantity with angular momentum $\boldsymbol{j} = \boldsymbol{l} + \boldsymbol{s}$. For example, a Ce^{3+} ion contains a single unpaired $4f$-electron in the state $4f^1$, with $l = 3$ and $s = 1/2$. Spin-orbit coupling gives rise to low-order multiplet with $j = 3 - 1/2 = 5/2$ with an associated magnetic moment $M = 2.64\mu_B$. In the crystal, this $2j + 1 = 6$ fold degeneracy is split by a crystalline electric field, then most of the ground state becomes doublet. One of the classic signatures of localized moments is a high-temperature Curie Weiss susceptibility, given by

$$\chi \propto \frac{1}{T - \theta} \tag{1.24}$$

θ is the Curie–Weiss temperature, a phenomenological scale that takes account of interactions between electron spins. The presence of such local moments inside a metal profoundly alters its properties. The physics of an isolated magnetic ion is described by the Kondo model

$$\mathcal{H} = \sum_{k\sigma} \epsilon c_{k\sigma}^{\dagger} c_{k\sigma} + J \psi^{\dagger}(0) \boldsymbol{\sigma} \psi(0) \cdot \boldsymbol{S}_f, \tag{1.25}$$

where $c_{k\sigma}^{\dagger}$ creates a conduction electron of energy. The conduction electron interacts with local moment via an antiferromagnetic interaction J. The effective strength of the interaction grows logarithmically, according to

$$J \to J(T) = J + 2J^2 \rho \ln \frac{D}{T} \tag{1.26}$$

where ρ is the density of states and D is band width.

1.2.3 Kondo Lattice and Doniach Phase Diagram

In heavy fermion material, containing a lattice of local moments, the Kondo effect develops coherence. In a single impurity, a Kondo singlet scatters electrons without conserving momentum, giving rise to a huge hump of resistivity at low temperatures. However, in a lattice, with translational symmetry, this same elastic scattering now conserves momentum, and this leads to coherent scattering of the Kondo singlets. In the simplest heavy fermion metals, this leads to a dramatic reduction in the resistivity at temperatures below the Kondo temperature.

As a typical Kondo lattice system, $CeCu_6$ is known [3]. At first glance, $CeCu_6$ is just Cu mixed with 14% Ce, but this alloy exhibits very interesting properties. By examining the effect of La substitution in this alloy, i.e. $La_xCe_{1-x}Cu6$, we can investigate how the single-ion Kondo becomes the heavy fermion system. Lanthanum is iso-electrical with Ce, but has no $4f$ electrons, so the alloy is in the dilute Kondo state in the $x \to 0$ limit. As a result, the resistivity shows a logarithmic temperature dependence. On the other hand, in the Ce dense region, the alloy becomes a coherent Kondo state and the resistivity shows T^2-behavior characteristic to the Fermi-liquid state.

$CeCu_6$ shows the following features inherent to the heavy fermion state [3]:

- Curie–Weiss susceptibility at high temperatures
- Van–Vleck susceptibility at low temperatures.
- Extremely high Sommerfeld coefficient $\gamma \sim 1000$ (mJ/mol·K^2)
- T^2 resistivity at low temperatures.

The competition between magnetism and the Kondo effect is well understood by the Doniach phase diagram [4] (Fig. 1.2). The Doniach model describes this phase diagram in terms of the competition between the indirect local moment exchange, represented by the RKKY interaction $T_{\mathrm{RKKY}} \sim J^2$, and the Kondo screening, represented by T_{K} $\exp(-1/J)$. Where, RKKY is an acronym for M.A. Ruderman, C. Kittel, T. Kasuya, and K. Yosida, and RKKY interaction is an interaction between

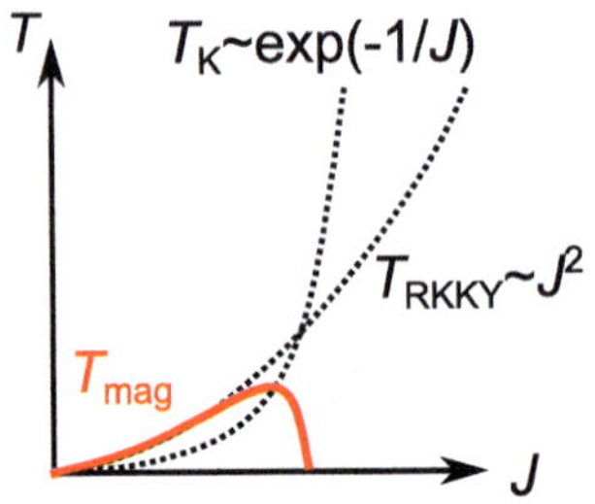

Fig. 1.2 Simple image for heavy fermion system. When J is small, f moment is localized, and magnetism occurs around T_{RKKY}. With increasing J, T_{K} drastically increases, and magnetism is suppressed

localized spins via conduction electrons. For small J, the RKKY exchange is dominant and the system is magnetically ordered at $T_{\mathrm{mag}} \simeq T_{\mathrm{RKKY}}$. At intermediate J, T_{RKKY} and T_{K} are of equal strength and magnetic ordering still occurs, but this time in increasingly shielded moments. Finally, as J is further increased, the magnetism is suppressed and $T_{\mathrm{mag}} = 0\,\mathrm{K}$ at the critical value (J_c). Around J_c, quantum critical behavior is observed, and sometimes superconductivity occurs. The superconductivity that occurs at these points is not BCS-like, but unconventional superconductivity is realized. Above J_c, the system shows the Fermi-liquid behavior with a characteristic energy scale $T_{\mathrm{FL}} \sim T_{\mathrm{K}}$.

1.3 U-Based Ferromagnetic Superconductors

1.3.1 Ferromagnetic Superconductors

Starting with the discovery of the superconducting UGe_2 in 2000 [5], a family of U-based ferromagnetic superconductors was discovered: the URhGe [6] and UCoGe [7]. The characteristic feature of these systems is the microscopic coexistence of ferromagnetism and superconductivity. The coexistence of ferromagnetism and superconductivity has been reported in other systems, such as $ErRh_4B_4$ [8] and $(Ce,Gd)Ru_2$ [9], but in those systems, the ferromagnetic electrons and the superconducting electrons are from different sites.

In 2000, the itinerant ferromagnet UGe_2 was discovered as the "true" coexistence of ferromagnetism and superconductivity under hydrostatic pressure ~1 GPa [5]. Subsequently, coexistence at ambient pressure was discovered in URhGe [6] and UCoGe [7]. So far, ferromagnetic superconductivity has only been discovered in U-based compounds. In these systems, U-5f electrons have been shown to be involved in both ferromagnetism and superconductivity. Another common feature is that ferromagnetism can be easily controlled by external magnetic fields and pressure, which implies that ferromagnetism is near the quantum critical point. Furthermore, it has been shown that the presence of this ferromagnetism is important for superconductivity. Therefore, these compounds are ideal systems for studying the relationship between ferromagnetic quantum criticality and superconductivity. Because of

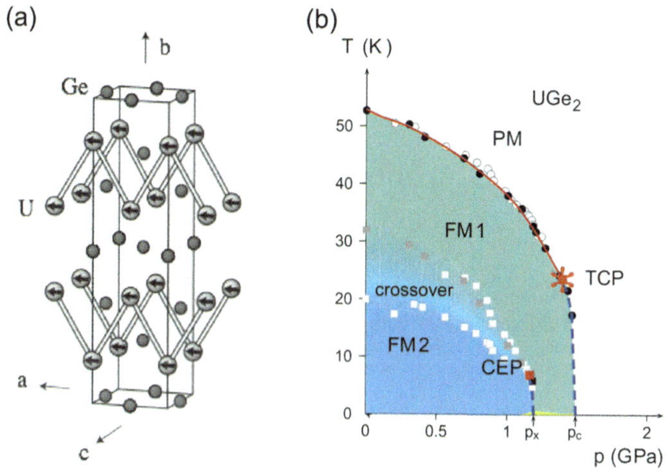

Fig. 1.3 **a** Crystal strucuture and **b** P–T phase diagram of UGe_2. These figures are reproduced with permission from Ref. [10] © (2019) The Physical Society of Japan

the ferromagnetism, spin-triplet pairing is expected in these systems, thus, U-based ferromagnetic SCs are the ideal platform to study spin-triplet superconductivity.

UGe_2

The crystal structure of UGe_2 is shown in Fig. 1.3 and belongs to an orthorhombic structure $Cmmm$ space group (#65, D_{2h}^{19}). UGe_2 shows ferromagnetism below $T_{Curie} = 52$ K with the ordered moment $M_0 = 1.48\ \mu_B$ per U atom along the a axis.

As shown in the phase diagram, the ferromagnetism in UGe_2 has two regions (FM1, FM2). The magnetic moment of each region has a different value and the transition from FM1 to FM2 is known as a 1st order FM transition. Superconductivity is observed only in the FM dome, and interestingly T_{SC} has the highest value at the border of transition from FM1 to FM2 ($P = 1.2$ GPa), suggesting that fluctuations related to the FM2 phase play an important role for the superconductivity.

The evidence for the microscopic coexistence of ferromagnetism and superconductivity was reported using NQR measurements [11]. Below the FM transition temperature, the ^{7}3Ge-NQR signal split owing to the spontaneous internal magnetic field. Further decreasing temperatures, SC transition occurs, and nuclear spin-lattice relaxation rate ($1/T_1$) quickly decreases below T_{SC} at the FM site. This result is direct evidence of the appearance of the superconducting gap inside the FM phase.

URhGe

URhGe crystallizes $Pnma$ space group (#62, D_{2h}^{16}), which belongs to a nonsymmorphic space group, as shown in Fig. 1.4. Different from UGe_2, superconductivity emerges even at ambient pressure, and T_{Curie} and T_{SC} are 9.5 K and 0.26 K, respectively. The ordered magnetic moment is around 0.4 μ_B along the c axis, indicating the strong itinerancy of this compound even in the FM state.

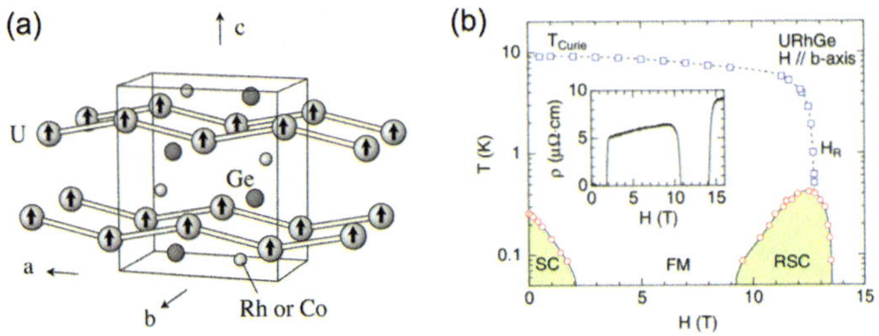

Fig. 1.4 **a** The crystal structure of URhGe, *Pnma* space group. **b** The *H*–*T* phase diagram of URhGe. These figures are reproduced with permission from Ref. [10] © (2019) The Physical Society of Japan

The most conspicuous feature in URhGe is reentrant superconductivity, i.e., the reappearance of a superconducting phase in a magnetic field as shown in Fig. 1.4. When a magnetic field is applied in the *b*-axis direction, the superconductivity is first destroyed at about 2 T. When the magnetic field is further increased, superconductivity is restored between 9 and 13 T. More surprisingly, this reentrant superconducting phase shows a higher transition temperature than the low-field one.

UCoGe

UCoGe is the isostructural compound of URhGe. T_{Curie} and T_{SC} is of about 3 K and 0.6 K, respectively. 0.6 K is the highest among the ferromagnetic superconductors. The magnetic moment M_0 is around 0.05 μ_{B} along the *c* axis and this value is smallest in these systems. A different point with URhGe is the H_{c2}. Unlike that of URhGe, H_{c2} in UCoGe exhibits enhanced rather than reentrant behavior. The H_{c2} of UCoGe is lowest in the *a* axis, the magnetic easy axis, at about 0.5 T. On the other hand, it exceeds 15 T in the *b* and *c* axes, and the field-boosted behavior is observed in the *b* axis direction.

1.3.2 Comparison of the Ferromagnetic Superconductors

There are many similarities and differences in the ferromagnetic superconductor. Here, I will introduce some important points related the present study.

Unusual H_{c2}: Field Reinforced Superconductivity

H_{c2} exhibits non-monotonic behavior in the three ferromagnetic superconductors in common. The *H*–*T* phase diagrams of three compounds are shown in Fig. 1.5. This direction is the *a* axis, the easy axis in UGe_2, and the *b* axis, the hard axis in URhGe and UCoGe. The enhancement of the superconductivity is strongly related with the ferromagnetic fluctuation revealed in UCoGe and URhGe using NMR technique [12, 13]. However, as mentioned above, the enhancement direction is not the same. The reason is still unresolved.

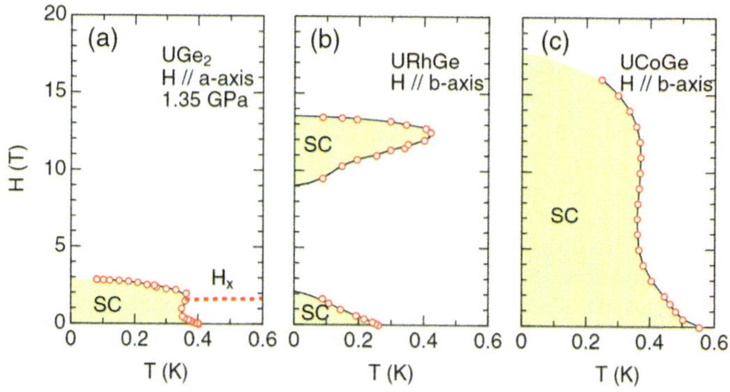

Fig. 1.5 The H–T phase diagrams of U-based ferromagnetic superconductors (**a** for UGe_2), **b** for URhGe, and **c** for UCoGe. These figures are reproduced with permission from Ref. [10] © (2019) The Physical Society of Japan

Scaling Law of T_{SC} and γ

The internal magnetic field H_{int} is much larger than the lower critical field H_{c1}, thus self-induced-vortex state exists at $H = 0$. The creation of self-induced vortex below T_{SC} will lead to the residual contribution $\gamma_0 \; H_{int}/H_{c2}$ to the linear T term. Figure 1.6 shows the SC anomaly of the three compounds; the SC jump at T_{SC} is directly related to the weakness of M_0. Even the residual γ term in the SC phase follows a $\sqrt{M_0}$ dependence.

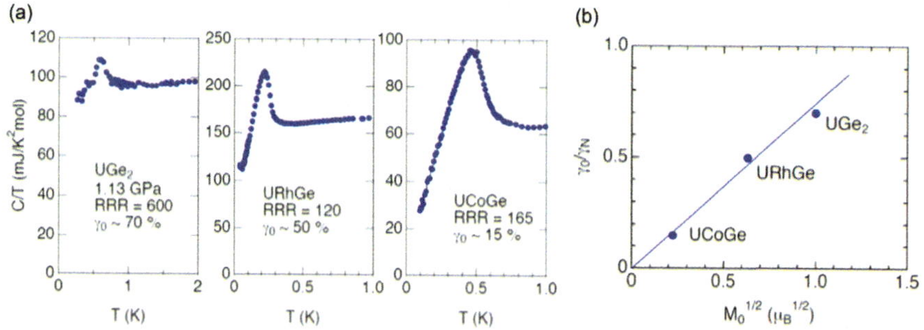

Fig. 1.6 **a** The temperature dependence of the specific heat C/T of UGe_2, URhGe, and UCoGe. **b** Relative variation of the residual value γ_0 normalized by γ_N as a function of $\sqrt{M_0}$. This graph proves the Volovik effect due to the self-induced vortex state. These figures are reproduced with permission from Ref. [10] © (2019) The Physical Society of Japan

References

1. Si Q, Steglich F (2010) Heavy fermions and quantum phase transitions. Science 329:1161–1166
2. Smith J, Kmetko E (1983) Magnetism or bonding: a nearly periodic table of transition elements. J Less Common Metals 90:83–88
3. Onuki Y, Shimizu Y, Komatsubara T (1984) Magnetic property of a New Kondo lattice intermetallic compound: CeCu6. J Phys Soc Jpn 53:1210–1213
4. Doniach S (1977) Phase diagram for the Kondo lattice. In: Parks RD (ed) Valence instabilities and related narrow-band phenomena. Springer, US, Boston, MA, pp 169–176
5. Saxena SS, Agarwal P, Ahilan K, Grosche FM, Haselwimmer RKW, Steiner MJ, Pugh E, Walker IR, Julian SR, Monthoux P, Lonzarich GG, Huxley A, Sheikin I, Braithwaite D, Flouquet J (2000) Superconductivity on the border of itinerant-electron ferromagnetism in UGe2. Nature 406:587–592
6. Aoki D, Huxley A, Ressouche E, Braithwaite D, Flouquet J, Brison J-P, Lhotel E, Paulsen C (2001) Coexistence of superconductivity and ferromagnetism in URhGe. Nature 413:613–616
7. Huy NT, Gasparini A, de Nijs DE, Huang Y, Klaasse JCP, Gortenmulder T, de Visser A, Hamann A, Görlach T, Löhneysen HV (2007) Superconductivity on the border of weak itinerant ferromagnetism in UCoGe. Phys Rev Lett 99:067006
8. Sinha SK, Crabtree GW, Hinks DG, Mook H (1982) Study of coexistence of ferromagnetism and superconductivity in single-crystal ErRh4B4. Phys Rev Lett 48:950–953
9. Kitaoka Y, Chang N-S, Ebisu T, Matsumura M, Asayama K, Kumagai K-I (1985) The coexistence of superconductivity and ferromagnetism in (Ce1 - xGdx)Ru2-self induced vortex state? J Phys Soc Jpn 54:1543–1551
10. Aoki D, Ishida K, Flouquet J (2019) Review of U-based ferromagnetic superconductors: comparison between UGe2, URhGe, and UCoGe. J Phys Soc Jpn 88:022001
11. Harada A, Kawasaki S, Kotegawa H, Kitaoka Y, Haga Y, Yamamoto E, Onuki Y, Itoh KM, Haller EE, Harima H (2005) Cooperative phenomenon of ferromagnetism and unconventional superconductivity in UGe2: A 73Ge-NQR study under pressure. J Phys Soc Jpn 74:2675–2678
12. Hattori T, Ihara Y, Nakai Y, Ishida K, Tada Y, Fujimoto S, Kawakami N, Osaki E, Deguchi K, Sato NK, Satoh I (2012) Superconductivity induced by longitudinal ferromagnetic fluctuations in UCoGe. Phys Rev Lett 108:066403
13. Tokunaga Y, Aoki D, Mayaffre H, Krämer S, Julien M-H, Berthier C, Horvatić M, Sakai H, Kambe S, Araki S (2015) Reentrant superconductivity driven by quantum Tricritical fluctuations in URhGe: evidence from 59Co NMR in URh0.9Co0.1Ge. Phys Rev Lett 114:216401

Chapter 2
UTe_2: Paradise for Condensed-Matter Scientists

2.1 Overview

The first report of superconductivity in UTe_2 was reported by Ran et al. in 2018 [1], and soon after reproduced by Aoki et al. [2]. At first, the SC transition temperature is around 1.6 K. They also confirm the magnetic susceptibility for three crystalline axes and H–T phase diagram for UTe_2. Because this behavior is very similar to the FMSCs, UTe_2 was considered as a member of the U-based FMSCs. The vital difference from U-based FMSCs is the absence of any magnetic order which was confirmed down to 50 mK by μSR measurement. Therefore, UTe_2 was considered to be located at the paramagnetic endpoint suggested by the series of FMSCs. This compound shows one of the richest phase diagrams in unconventional superconductors and possesses many superconducting phases. Resulting, UTe_2 is the paradise for the condensed-matter scientist.

2.2 Basic Properties

2.2.1 Sample Preparation

The first single crystal of UTe_2 was reported by Stöwe by chemical vapor transport (CVT) method with $TeBr_4$ as a transport agent [3]. The detailed electronic properties were first reported by Ikeda et al. [4]. They made single crystalline UTe_2 by CVT with I as a transport agent. They measured the resistivity, susceptibility, and specific heat, and a similar value was reported. However, there was no sign of superconductivity. The first superconductivity was reported by Ran et al. [1]. They also made single crystals by the CVT method, but the starting ratio was not stoichiometric, and the ratio of U:Te was 2:3. Later, Aoki et al. [5] and Rosa et al. [6] clarified the conditions of the

K. Kinjo, *Nuclear Magnetic Resonance Study on Multiple Superconducting Phases in UTe2*, Springer Theses, https://doi.org/10.1007/978-981-97-8646-6_2

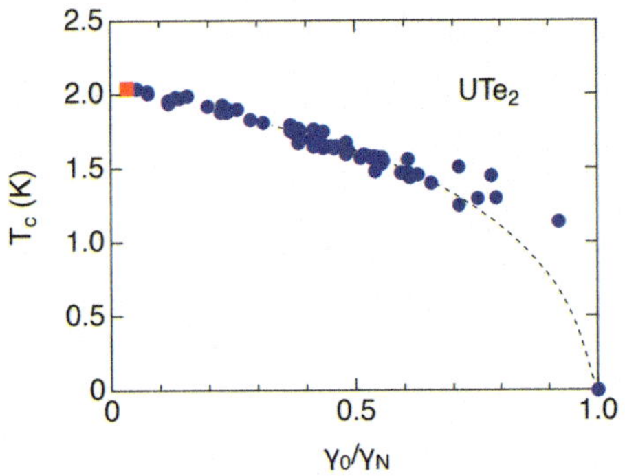

Fig. 2.1 T_c versus residual γ-value (γ_0) scaled by that in the normal state (γ_N) in different quality samples. The red square is the sample made by the flux method. This figure is reproduced with permission from Ref. [9] © (2022) The Physical Society of Japan

CVT method more clearly, and the transition temperature was raised to about 2.0 K. It was also found that the higher the transition temperature, the smaller the residual specific heat term becomes (Fig. 2.1). Then, after examining various samples, Haga et al. found that a slight deficiency of U in the sample caused poor superconductivity [7]. Finally, growth by the flux method in a reducing atmosphere, which is less likely to produce U deficiency, was reported by Sakai et al. in the middle of 2022, and the transition temperature increased to 2.1 K [8]. At present, this is considered to be the best sample.

2.2.2 *Crystal Structure*

As shown in Fig. 2.2, UTe_2 crystallizes in an orthorhombic $Immm$ space group (#71), which has D_{2h} point group symmetry. In UTe_2, as shown in the figure, two uranium atoms, connecting to the c-axis, are coupled as ladder-like along the a axis, and this ladder stacks along the c axis. This structure is symmorphic and different from URhGe and UCoGe. All atomic positions are listed in Table 2.1. There are two crystallographically inequivalent Te sites in UTe_2, $4j$ and $4h$, sites with point symmetries $mm2$ and $m2m$, respectively. Due to this structure, we can obtain two ^{125}Te-NMR signals as shown later.

2.2.3 *Fermi Surface*

Theoretical Calculation

From conventional LDA calculation, UTe_2 was shown to be an insulator with a small gap of about 13 meV [2]. Thus, Ishizuka et al. took into account the electron correlations from first-principles so-called GGA+U methods [11], where GGA is

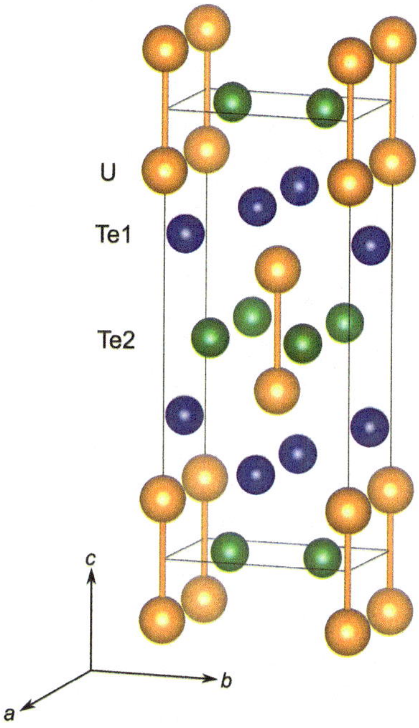

Fig. 2.2 Schematic view of the body-centered orthorhombic crystal structure of UTe_2. This figure is drawn by VESTA [10]

Table 2.1 Site symmetry of each atom

Atom	Wycoff letter	Symmetry
U	4*i*	*mm*2
Te1	4*j*	*mm*2
Te2	4*j*	*m*2*m*

"Generalized gradient approximation" and +U is considering the on-site Coulomb repulsion. As shown in the inset of Fig. 2.3b, the insulating gap closes and becomes metallic.

dHvA Measurements

The dHvA signal was not observed for three years after the first discovery of superconducting UTe_2 in 2019. Higher-quality single crystals with long mean free path were successfully grown after improving the sample growth method, and the dHvA signal was observed for the first time [12]. Figure 2.3 shows the angular dependence of the dHvA frequencies for the field in the *ac* plane. Two samples show almost the same dHvA signals. With increasing the field angle, branch1 slightly decreases first, then increases up to 50°. Branchβ increases continuously up to 45°. Branch 2 also increases with the field angle and shows a sharp increase around 60°. No dHvA

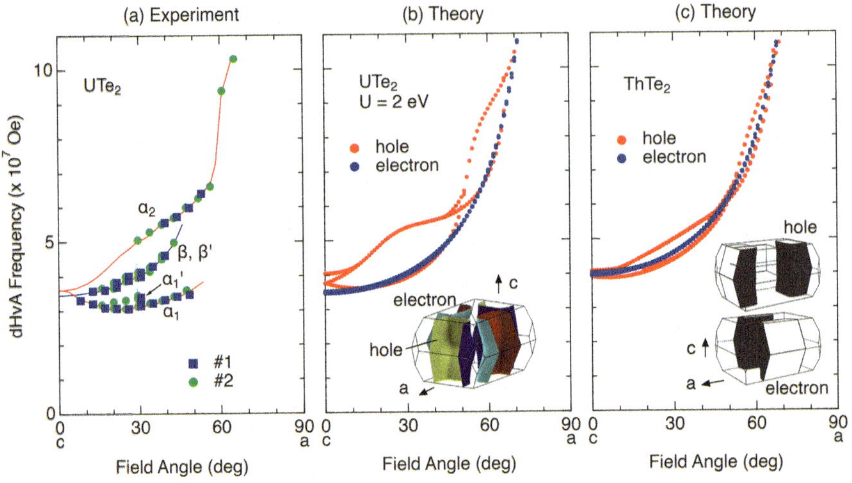

Fig. 2.3 **a** Angular dependence of the dHvA frequency in UTe_2. The square and circle represent the different samples. The lines are guides to the eyes. Panels **b**, **c** show theoretical angular dependence of the dHvA frequency from the band calculations based on the GGA+U (U = 2 eV) method as mentioned above, and the LDA calculation in $ThTe_2$, assuming the tetravalent U atom with the localized $5f^2$ configuration in UTe_2. The corresponding Fermi surfaces are depicted. These figures are reproduced with permission from Ref. [9] © (2022) The Physical Society of Japan

signal was detected around the $H \parallel a$ axis. These features are in good agreement with the theoretical calculation with $U = 2$ eV as shown in the inset of the figure.

2.2.4 *Normal State Properties*

Resistivity

The electrical resistivity strongly depends on the current direction, as shown in Fig. 2.4a. While the resistivity shows a broad maximum at around 50 K and decreases on cooling due to the formation of Kondo coherence for $j \parallel a$ and b, the for $j \parallel c$ shows a minimum around 40 K and sharp peak at 15 K. At low temperatures, all resistivities follow the T^2 behavior, indicating the formation of the heavy-fermion state.

Magnetic Susceptibility

The magnetic susceptibility χ of UTe_2 is shown in Fig. 2.4b for the three crystalline axes a, b, and c. For all directions, the susceptibility follows a Curie-Weiss behavior with $\chi = C/(T - \theta)$ at high temperatures above $\sim$ 150 K. For all crystalline axes, the Weiss temperature θ is around -100 K, indicating the antiferromagnetic correlations at high temperatures. This seems inconsistent with the similarity to the FMSCs, but, the negative Weiss temperature at high temperatures was also observed in UCoGe

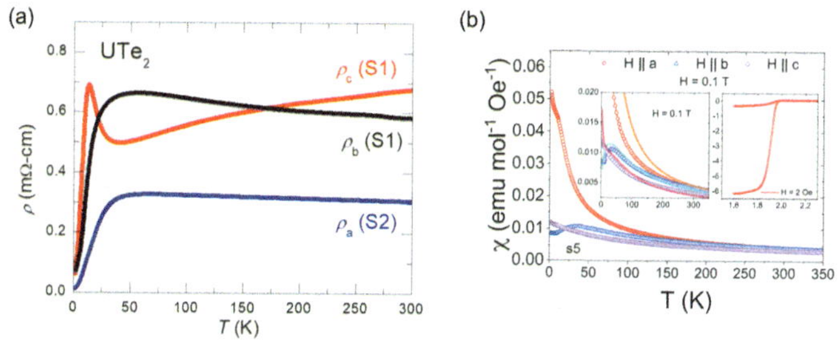

Fig. 2.4 **a** The temperature dependence of the electrical resistivity of UTe_2 for a current applied along three crystalline axes. **b** The temperature dependence of the magnetic susceptibility of UTe_2. The inset shows the field dependence of the magnetization measured at 1.8 K. Panel **a** was reproduced with permission from Ref. [14] © (2022) The American Physical Society. Panel **b** was reproduced with CC BY license from Ref. [6]

and URhGe. These results indicate the difference in magnetic character between the high-temperature and low-temperature regions. The estimated effective moment of the U atom at high-temperature region is around 3.4 μ_B. This value is very close to the free ion values 3.58 μ_B for U^{4+} ($5f^2$) or 3.64 μ_B for U^{3+} ($5f^3$). On cooling, χ_a strongly increases with stronger than the Curie Weiss behavior below 100 K, and starts to saturate around 20 K. Below 12 K, χ_a again increases and diverges towards $T \to 0$, however, this is not the intrinsic effect but due to the small U-deficiency effect[1].

In difference to the *a* axis, χ_b follows the Curie-Weiss behavior down to 50 K and shows a broad maximum at $T_{\chi_{max}} \sim 35$ K. Such a behavior has been often observed in paramagnetic systems along the easy magnetization axis due to the formation of the Heavy fermion state [13]. As shown in later, this $T_{\chi_{max}}$ smoothly connects the metamagnetic transition line. This also supports the formation of a heavy fermion state. Above 25 K, χ_c is smallest, however, with decreasing temperature χ_c increases, and thus, below 25 K, the *b* axis becomes the hardest axis. At 1.8 K, the anisotropy $M_a/M_b \sim 5.5$ is not so strong Ising-type, as in UCoGe $M_a/M_b \sim 8$. Therefore, UTe_2 has weak Ising-type susceptibility.

2.2.5 *Metamagnetic Transition*

For $H \parallel b$, a metamagnetic transition occurs at $H_m \sim 35$ T. Although a metamagnetic transition occurs for the magnetic field along the easy axis in the usual nearly ferromagnetic system, the metamagnetic transition occurs for the magnetic hard axis in UTe_2. As shown in Fig. 2.5, below 8 K, magnetization shows a strong jump at around 35 T with a huge jump of $\Delta M_b \sim 0.5$ μ_B and clear hysteresis. These results indicate the first-order phase transition at this point. Above a critical endpoint $T_{CEP} \sim 8$ K,

[1] From private communication

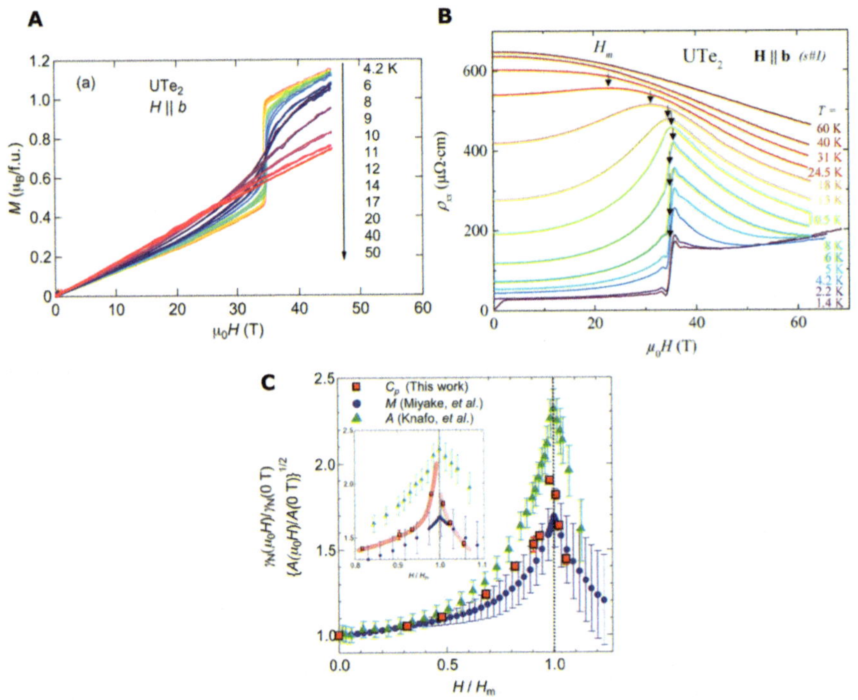

Fig. 2.5 **a** The field dependence of the magnetization for $H \parallel b$ at different temperatures. **b** Magnetoresistivity for $H \parallel b$ at different temperatures. **c** Magnetic field dependence of γ_N normalized to $\gamma_N(H = 0)$ as function of field normalized to H_m. The red squares are from specific heat measurements in a pulsed magnetic field, the blue circles from magnetization measurements using the Maxwell relation, and the green triangles from the A coefficient of the resistivity. The inset gives a zoom on the field range near H_m. These figures are reproduced with CC BY license from Ref. [5]

this first-order transition changes into a crossover, and M_b shows an inflection point as a function of H_b. Magnetoresistivity measurements for $H \parallel b$ also show a huge jump accompanied by a magnetization jump, as shown in Fig. 2.5. Magnetoresistivity also shows the change of transition character.

The electronic correlation increases with approaching H_m. The A coefficient, which is the coefficient of the square of the temperature of the electrical resistivity ($\rho = \rho_0 + AT^2$), is strongly enhanced up to H_m. The Sommerfelt coefficient γ is also enhanced around H_m. γ is measured by specific heat and magnetization measurements using Maxwell relation[2].

[2] From thermodynamics, the relation between magnetization M and entropy S is obtained from the Maxwell relation: $(\partial S/\partial H)_T = (\partial M/\partial T)_H$. In addition, at low temperatures, C/T is almost constant. Therefore, $S = \gamma T$ in UTe_2 is hold up. As a results, we can detect the H dependence of γ as follows: $(\partial \gamma/\partial H)_T = (\partial^2 M/\partial T^2)_H$.

2.2.6 Spin Fluctuations

The nature of the spin fluctuations in UTe_2 has been investigated by means of ^{125}Te-NMR and inelastic neutron scattering (INS).

Ferromagnetic Fluctuation: From NMR Measurements

As written in Sect. 2.3, NMR can measure magnetic or electronic fluctuations. In the case of ^{125}Te-NMR, since ^{125}Te nuclei have spin 1/2, only magnetic fluctuation is detected. Figure 2.6a shows the temperature dependence of the NMR spin-lattice relaxation rate $(1/T_1T)_\alpha$ measured in a single crystal for the field along all three crystalline axis direction ($\alpha = a, b$, and c). In contrast with temperature independent behavior of $(1/T_1T)_a$, $(1/T_1T)_{b,c}$ strongly increases on cooling.

As mentioned in the next section, $1/T_1T$ reflects the fluctuations along the direction which are perpendicular to the quantization axis, *i.e.* magnetic field direction. The value of $1/T_1T$ in the three crystalline axes can be described as follows using the α-axis fluctuation R_α: $(1/T_1T)_\alpha = R_\beta + R_\gamma$, where α, β, γ is the different three crystalline axes. The results are shown in Fig. 2.6. UTe_2 exhibits a strong anisotropy for the spin fluctuations, *i.e.*, $R_a \gg R_c > R_b$.

Incommensurate Antiferromagnetic Fluctuation: INS Measurements

Inelastic neutron scattering (INS) measurements can reveal magnetic excitation. A first INS experiment has been performed at the Cold Neutron Chopper Spectrometer at Oak Ridge National Laboratory [16] Surprisingly, antiferromagnetic fluctuations at the incommensurate wavevectors $\mathbf{k_1} = (0, 0.57, 0)$ and $\mathbf{k}_2 = (0, 0.43, 0)$ (in reciprocal lattice units) which extends at least up to 2.6meV as shown in Fig. 2.7 was observed. The obtained incommensurate wave vectors are very close to the commensurate antiferromagnetic vector $(0, 0.5, 0)$. The emergence of these antiferromagnetic fluctuations is explained by electronic structure calculations and it has

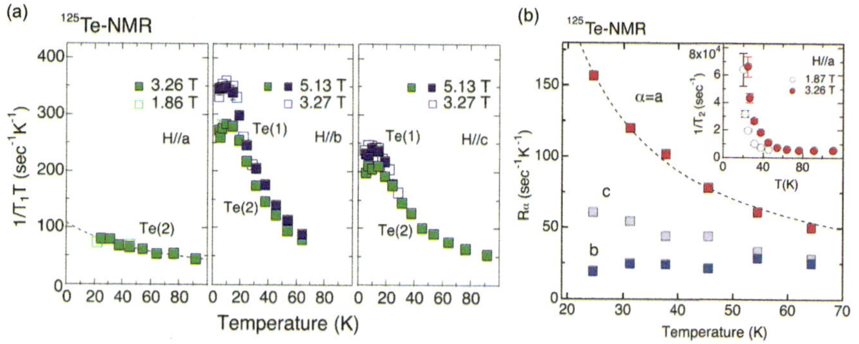

Fig. 2.6 **a** The temperature dependence of $(1/T_1T)$ for all three crystalline axes. **b** Spin fluctuations along each crystalline axis calculated using $(1/T_1T)$. The inset shows the temperature dependence of $1/T_2$ for the field along the a-axis. These figures are reproduced with permission from Ref. [15] © (2019) The Physical Society of Japan

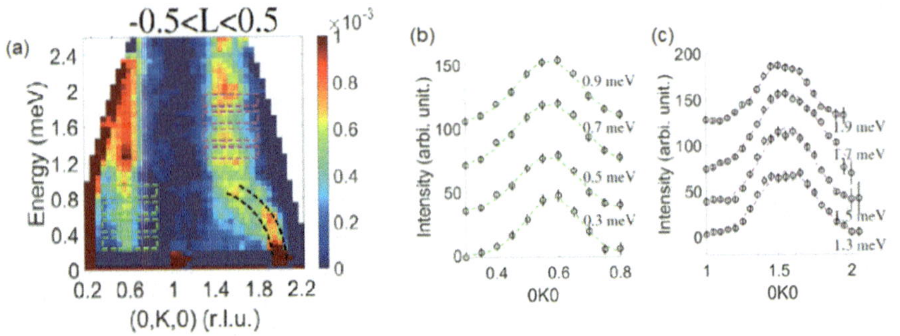

Fig. 2.7 **a** Energy dependence of the magnetic excitations measured along the [0, K, 0] direction revealing two excitations at $k_1 = (0, 0.57, 0)$ and $k_2 = (0, 1.57, 0)$ integrated between $-0.5 \leq L \leq 0.5$. **b** [**c**] 1D cut along the K direction at different energies through the $k_1[k_2]$. These figures are reproduced with permission from Ref. [16] © (2020) The American Physical Society

been concluded that the RKKY interaction between the $5f$ moments drives the antiferromagnetic fluctuations.

Neutron scattering measurements only find incommensurate magnetic fluctuations near an antiferromagnetic wave vector, in contrast to the initial expectation of ferromagnetic correlation. These antiferromagnetic correlations involve ferromagnetic correlations along the a axis in the ladder direction, with antiferromagnetic inter-ladder coupling along the b axis, and a quasi-magnetic decoupling along the c axis.

2.3 Superconductivity

The first observation of superconductivity in UTe_2 by S. Ran *et. al.* using the chemical vapor transport method, and its SC critical temperature (T_c) is around 1.6 K [1]. After many experimental signs of progress, H. Sakai et al., found a new root to get more high-quality single crystals with the flux method, and T_c increases up to 2.1 K. However, only a few experiments were done using such a clean sample at present, therefore main results shown below are mainly using 1.6–1.8 K samples.

2.3.1 *Field Boosted Superconductivity: Enhancement of H_{c2}*

First, I will introduce the upper critical field of UTe_2. As shown in Fig. 2.8, the three H_{c2}s for principal axes exceed conventional Pauli limit $H_P \sim 1.84T_c[T/K] \sim 3$ T, supporting the spin-triplet scenario [1, 2, 17]. In addition, by applying H greater than 15 T to the b axis (hard axis), T_c starts to increase, and the L-shape H-T SC phase appears as shown in Fig. 2.8. This behavior is similar to the S-shape SC phase in

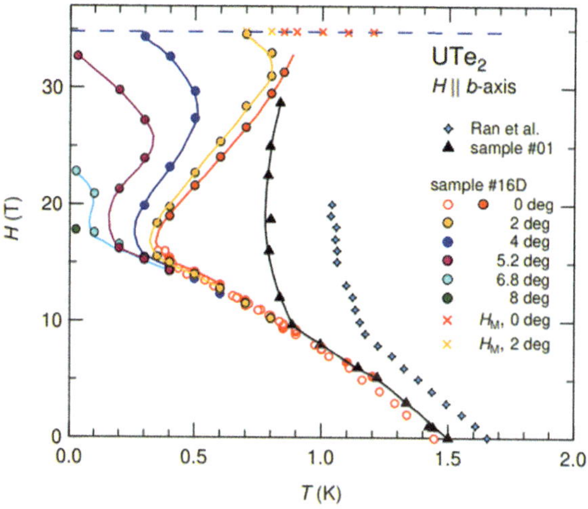

Fig. 2.8 Angular dependence of H_{c2} from the b up to 8° turned to the a axis. This figure is reproduced with permission from Ref. [17] © (2019) The Physical Society of Japan

UCoGe and the reentrant SC phase in URhGe. Since the H-T phase diagram is similar to those of the U-based ferromagnetic superconductors, UTe_2 is also considered to be one of the candidates for a spin-triplet superconductor, although any magnetic order has not been found.

As mentioned in the previous section, the superconductivity for $H \parallel b$ is suddenly destroyed with metamagnetic transition, and the magnetic moment of 0.1 μ_B arises [18]. The phase is called a fully polarized (FP) phase and the origin is still unclear, but the metamagnetic transition is considered to be the breakdown of the Kondo state. Surprisingly, by applying the tilted field of 20°–40° from b to c axis, superconductivity reappears in the FP phase [19, 20]. Figure 2.9 is the H angle–H

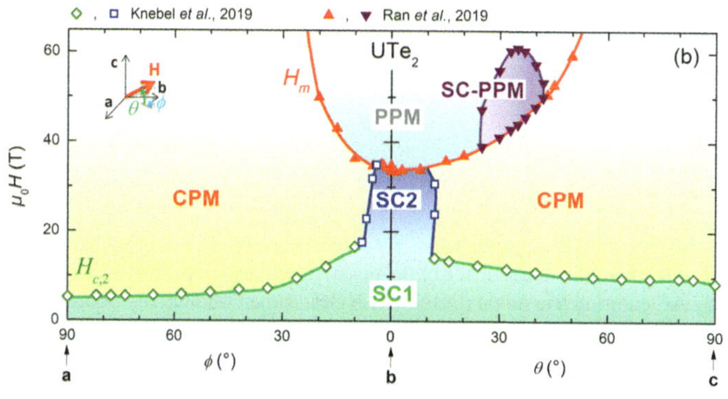

Fig. 2.9 The H angle–H phase diagram of UTe_2. SC in FP (SC-PPM) appears where the magnetic field magnitude is above 40 T and the magnetic field angle is tilted about 30° from the b axis to the c axis. This figure is reproduced with CC 4.0 licence from Ref. [22]

phase diagram to see the such superconducting state. Around the H_m, the divergence of A coefficient of the resistivity and the Sommerfeld coefficient was reported [18, 21]. These results indicate the enhancement of effective mass m^* linked to magnetic instability and the possible relation to the high-H SC phase.

2.4 Pressure Effect

In various superconductors, their SC properties varies with pressure. Pressure effects have also been intensively studied in UTe_2 with such motivation. Figure 2.10a shows the temperature dependence of the C/T obtained from the AC calorimetry measurement on UTe_2. It shows the clear successive transitions at 0.31 and 0.47 GPa, indicating multiple SC phases induced by applying pressure [23]. Figure 2.10b shows the temperature dependence of the resistivity at various pressure up to 3.2 GPa. Below 1.3GPa, resistivity suddenly drops at several temperatures. On the other hand, above 1.6 GPa, resistivity shows hump-like behavior, suggesting magnetic ordering. The P-T phase diagram was summarized in Fig. 2.10c. For now, all studies of the high-pressure phase diagram suggest that the transition from the SC state to the magnetically ordered state is first order, as indicated in Fig. 2.10c. There is no direct evidence to determine the magnetic order, however, recent experiments conclude the possibility of AFM ground state above P_c. The characteristic temperature of 15 K anomaly decreases linearly with applying pressure down to around 2 K close to P_c [24].

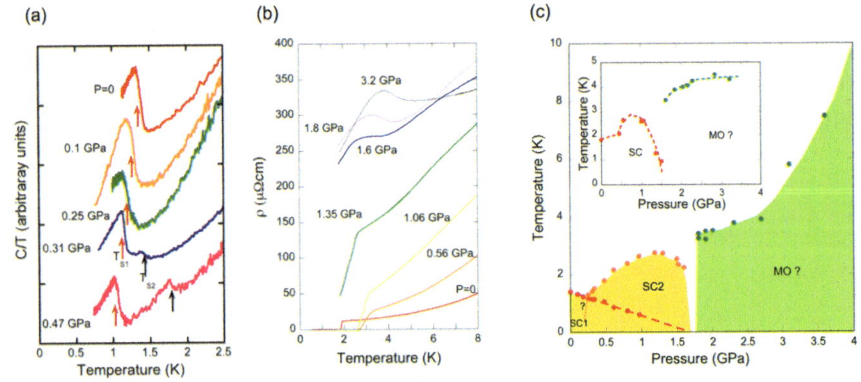

Fig. 2.10 **a** The temperature dependence of C/T measured by AC-calorimetric method at several pressures. **b** The temperature dependence of the resistivity at several pressures. **c** P-T phase diagram determined by AC calorimetric method. Below 1.6 GPa, superconductivity occurs. Above 1.6 GPa, a magnetically ordered state was found. This figure is reproduced with CC 4.0 licence from Ref. [23]

2.4.1 Field Effects for SC Under Pressure

The H–T phase diagrams of UTe_2 under several pressures are shown in Fig. 2.11 Resistivity and AC calorimetry studies under pressure have confirmed the occurrence of multiple phase transitions induced by magnetic fields, confirming a superconducting transition [25]. For $H \parallel a$, a kink in the temperature dependence of H_{c2} was observed in the electrical resistivity at low temperatures at $P = 0.5$ GPa [25]. This was quickly confirmed by AC calorimetry and the presence of multiple superconducting phases below H_{c2} was established. For $H \parallel b$, it is shown that H_{c2} is limited by H_m and that the pressure dependence of H_m follows that of $T_{\chi_{\max}}$ [25]. The enhancement of superconductivity by magnetic field disappears above 1 GPa. For $H \parallel c$, the H_{c2} has been determined by magnetoresistance and calorimetry. As the

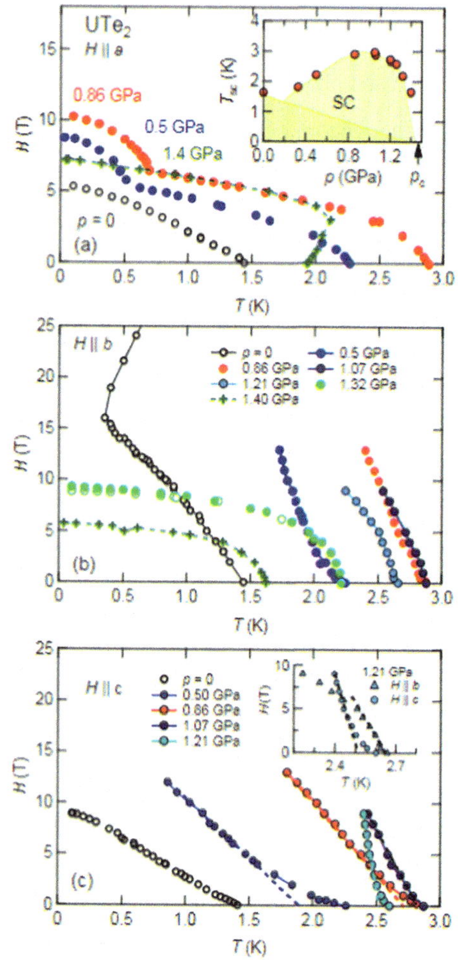

Fig. 2.11 The upper critical field of UTe_2 at several pressures. These figures are reproduced with permission from Ref. [25] © (2020) The Physical Society of Japan

critical pressure p_c is approached, a strong increase in $\partial H_{c2}/\partial T$ and a nearly vertical slope are observed [23]. Near the p_c ($P = 1.45$ GPa), the long-range antiferromagnetic order (MO) appears to limit the superconducting region to a narrow region at high field and low temperature. At p_c, no superconductivity is observed at zero fields, but field-reentrant superconductivity seems to survive just above the critical field of the magnetic order. Finally, at 1.61 GPa, no more traces of superconductivity are observed [26].

2.4.2 *Possible AFM Ordering*

As mentioned above, the order parameter of MO phase is still unknown, however, there is some circumstantial evidence suggesting AFM order. The experimental facts are listed below.

1. Hump like behavior in T-dependence of the resistivity
2. Closed phase diagram for all magnetic field directions

1. When incommensurate spin density wave order (Metalic AFM-like state) occurs, a hump-like anomaly is often observed, *i.e.* a behavior in which the electrical resistance increases due to the increase of magnetic scattering and then decreases due to the effect of temperature drop. On the other hand, the resistivity decrease with a kink-like anomaly is often observed in ferromagnetic transitions.
2. In the FM state, a magnetic moment appears in a certain axial direction. This means that under a magnetic field in that axial direction, FM transition becomes a crossover and its phase diagram is no longer closed. In the MO phase of UTe_2, the phase diagram is closed to magnetic fields in the three crystalline principle directions, so it is not FM, but likely AFM order.

2.5 Purpose of This Book

In the previous sections, I reviewed the normal- and SC-state properties of UTe_2. To investigate the SC properties of UTe_2 more directly, unveiling the SC spin state is crucial. In my study, we measured a single crystal sample of UTe_2, T_c of which were $\sim$1.6 K. The sample is called the "1.6 K" sample in this book. The main studies in this book are as follows:

1. Superconducting properties of UTe_2 **of LFSC state: Sect.** 5.1
In this section, we prove that UTe_2 is a spin-triplet superconductor. In spin-triplet superconductors, anisotropy appears in the reduction of the spin susceptibility, as shown in Chap. 1. We have performed NMR measurements in single crystalline

UTe_2 for three principal axes. Our results unveil that the spins of the SC pairs are oriented along the a-axis at least at $H = 0.56$ T in the 1.6 K sample.

2. The magnetic field rotates the superconducting spin: Sect. 5.2
In this section, we show that the superconducting spins of UTe_2 are rotated by the magnetic field. As mentioned above, the spin-triplet superconductivity has spin degrees of freedom, so the superconducting spins are theoretically considered to rotate with a magnetic field, but this has not been reported. Our NMR measurements up to 24 T observed distinct superconducting spin rotation. This clearly shows th at UTe_2 changes its superconducting characters depending on the magnetic field.

3. Pressure also rotates the superconducting spin: Sects. 5.3, 5.4
In these sections, we show that the superconducting properties of UTe_2 also vary with pressure, and propose an analogy between superconductivity under pressure and superconductivity induced by a magnetic field.

Section 5.3 **Drastic change in magnetic anisotropy in** UTe_2 **under pressure**
First, we measure the electronic state of UTe_2 under pressure. Since there are only a limited number of methods to measure physical properties under pressure, NMR measurement is very important because it is a non-contact measurement. Our NMR measurements show that the electronic state changes dramatically below and above the critical pressure (P_c). This indicates that superconductivity occurs only in the heavy-electron state of UTe_2, and not in the magnetically ordered state. This is in contrast with the case of FMSCs.

Section. 5.4 **Superconducting spin rotation in multiple superconducting phases of** UTe_2 **under pressrue**
In this section, we show that the superconducting spins of UTe_2 are also rotated by the pressure accompanied by phase transition. As mentioned above, UTe_2 has multiple SC phases under pressure and two SC phases appear in the intermediate pressure region on cooling. NMR measurements in the region showed that the spin susceptibility in the SC2 phase in Fig. 2.10 does not change for $H \parallel b$ and that the spin susceptibility changes in the low-temperature region. This indicates that the superconducting spins are rotating in the multiple SC phases. Furthermore, we show that this state is very similar to the high-field state, and the similarity allows a general explanation for the phase diagram in UTe_2.

A brief introduction to NMR measurements is given in Chap. 3, and the experimental settings are presented in Chap. 4. At the last, I will give the conclusion of this book and suggestions for future work in Chap. 6.

References

1. Ran S, Eckberg C, Ding QP, Furukawa Y, Metz T, Saha SR, Liu IL, Zic M, Kim H, Paglione J, Butch NP (2019) Nearly ferromagnetic spin-triplet superconductivity. Science 365:684–687
2. Aoki D, Nakamura A, Honda F, Li D, Homma Y, Shimizu Y, Sato YJ, Knebel G, Brison J-P, Pourret A, Braithwaite D, Lapertot G, Niu Q, Vališka M, Harima H, Flouquet J (2019) Unconventional superconductivity in heavy fermion UTe2. J Phys Soci Jpn 88:043702
3. Stowe K (1996) Contributions to the crystal chemistry of Uranium Tellurides: III. Temperature dependent structural investigations on Uranium Ditelluride. J Solid State Chem 127:202–210
4. Ikeda S, Sakai H, Aoki D, Homma Y, Yamamoto E, Nakamura A, Shiokawa Y, Haga Y, Onuki Y (2006) Single crystal growth and magnetic properties of UTe2. J Phys Soc Jpn 75:116–118
5. Aoki D, Brison J-P, Flouquet J, Ishida K, Knebel G, Tokunaga Y, Yanase Y (2022) Unconventional superconductivity in UTe2. J Phys: Condens Matter 34:243002
6. Rosa PFS, Weiland A, Fender SS, Scott BL, Ronning F, Thompson JD, Bauer ED, Thomas SM (2022) Single thermodynamic transition at 2 K in superconducting UTe2 single crystals. Commun Mater 3:33
7. Haga Y, Opletal P, Tokiwa Y, Yamamoto E, Tokunaga Y, Kambe S, Sakai H (2022) Effect of uranium deficiency on normal and superconducting properties in unconventional superconductor UTe2. J Phys: Condens Matter 34:175601
8. Sakai H, Opletal P, Tokiwa Y, Yamamoto E, Tokunaga Y, Kambe S, Haga Y (2022) Single crystal growth of superconducting UTe2 by molten salt flux method. Phys Rev Mater 6:073401
9. Aoki D, Ishida K, Flouquet J (2019) Review of U-based ferromagnetic superconductors: comparison between UGe2, URhGe, and UCoGe. J Phys Soc Jpn 88:022001
10. Momma K, Izumi F (2011) VESTA3 for three-dimensional visualization of crystal, volumetric and morphology data. J Appl Crystallogr 44:1272–1276
11. Ishizuka J, Sumita S, Daido A, Yanase Y (2019) Insulator-metal transition and topological superconductivity in UTe2 from a first-principles calculation. Phys Rev Lett 123:217001
12. Aoki D, Sakai H, Opletal P, Tokiwa Y, Ishizuka J, Yanase Y, Harima H, Nakamura A, Li D, Homma Y, Shimizu Y, Knebel G, Flouquet J, Haga Y (2022) First observation of the de Haas-van Alphen effect and fermi surfaces in the unconventional superconductor UTe2. J Phys Soc Jpn 91:083704
13. Kinjo K, Fujibayashi H, Nakamine G, Kitagawa S, Ishida K, Tokunaga Y, Sakai H, Kambe S, Nakamura A, Shimizu Y, Homma Y, Li D, Honda F, Aoki D (2022) Drastic change in magnetic anisotropy of UTe2 under pressure revealed by 125Te-NMR. Phys Rev B 105
14. Eo YS, Liu S, Saha SR, Kim H, Ran S, Horn JA, Hodovanets H, Collini J, Metz T, Fuhrman WT, Nevidomskyy AH, Denlinger JD, Butch NP, Fuhrer MS, Wray LA, Paglione J (2022) c-axis transport in UTe2: evidence of three dimensional conductivity component. Phys Rev B 106:L060505
15. Tokunaga Y, Sakai H, Kambe S, Hattori T, Higa N, Nakamine G, Kitagawa S, Ishida K, Nakamura A, Shimizu Y, Homma Y, Li D, Honda F, Aoki D (2019) 125Te NMR study on a single crystal of heavy fermion superconductor UTe2. J Phys Soc Jpn 88:073701
16. Duan C, Sasmal K, Maple MB, Podlesnyak A, Zhu JX, Si Q, Dai P (2020) Incommensurate spin fluctuations in the spin-triplet superconductor candidate UTe2. Phys Rev Lett 125:237003
17. Knebel G, Knafo W, Pourret A, Niu Q, Vališka M, Braithwaite D, Lapertot G, Nardone M, Zitouni A, Mishra S, Sheikin I, Seyfarth G, Brison J-P, Aoki D, Flouquet J (2019) Field-reentrant superconductivity close to a metamagnetic transition in the heavy-fermion superconductor UTe2. J Phys Soc Jpn 88:063707
18. Miyake A, Shimizu Y, Sato YJ, Li D, Nakamura A, Homma Y, Honda F, Flouquet J, Tokunaga M, Aoki D (2019) Metamagnetic transition in heavy fermion superconductor UTe2. J Phys Soc Jpn 88:063706
19. Ran S, Liu I-L, Eo YS, Campbell DJ, Neves PM, Fuhrman WT, Saha SR, Eckberg C, Kim H, Graf D, Balakirev F, Singleton J, Paglione J, Butch NP (2019) Extreme magnetic field-boosted superconductivity. Nat Phys 15:1250–1254

20. Knafo W, Nardone M, Vališka M, Zitouni A, Lapertot G, Aoki D, Knebel G, Braithwaite D (2021) Comparison of two superconducting phases induced by a magnetic field in UTe2. Commun Phys 4:40
21. Imajo S, Kohama Y, Miyake A, Dong C, Tokunaga M, Flouquet J, Kindo K, Aoki D (2019) Thermodynamic investigation of metamagnetism in pulsed high magnetic fields on heavy fermion superconductor UTe2. J Phys Soc Jpn 88:083705
22. Knafo W, Vališka M, Braithwaite D, Lapertot G, Knebel G, Pourret A, Brison JP, Flouquet J, Aoki D (2019) Magnetic-field-induced phenomena in the paramagnetic superconductor UTe2. J Phys Soc Jpn 88:063705
23. Braithwaite D, Vališka M, Knebel G, Lapertot G, Brison JP, Pourret A, Zhitomirsky ME, Flouquet J, Honda F, Aoki D (2019) Multiple superconducting phases in a nearly ferromagnetic system. Commun Phys 2:147
24. Ran S, Kim H, Liu I-L, Saha SR, Hayes I, Metz T, Eo YS, Paglione J, Butch NP (2020) Enhancement and reentrance of spin triplet superconductivity in UTe2 under pressure. Phys Rev B 101:140503
25. Knebel G, Kimata M, Vališka M, Honda F, Li D, Braithwaite D, Lapertot G, Knafo W, Pourret A, Sato YJ, Shimizu Y, Kihara T, Brison J-P, Flouquet J, Aoki D (2020) Anisotropy of the upper critical field in the heavy-fermion superconductor UTe2 under pressure. J Phys Soc Jpn 89:53707
26. Aoki D, Kimata M, Sato YJ, Knebel G, Honda F, Nakamura A, Li D, Homma Y, Shimizu Y, Knafo W, Braithwaite D, Vališka M, Pourret A, Brison J-P, Flouquet J (2021) Field-induced superconductivity near the superconducting critical pressure in UTe2. J Phys Soc Jpn 90:074705

Chapter 3
Nuclear Magnetic Resonance

3.1 Principle of NMR

3.1.1 *Nuclear Spin Hamiltonian*

Some atomic nuclei possess a nuclear magnetic moment, denoted as $\boldsymbol{\mu}_{\mathrm{n}}$. This moment interacts with external magnetic fields and other magnetic entities. The nuclear magnetic moment $\boldsymbol{\mu}\mathrm{n}$ can be expressed as $\boldsymbol{\mu}\mathrm{n} = \gamma_{\mathrm{n}}\hbar\boldsymbol{I}$, where γ_{n} represents the nuclear gyromagnetic ratio and $\boldsymbol{I}$ corresponds to the nuclear spin.

An external magnetic field H removes the degeneracy of nuclear spins, causing nuclear energy levels to be separated by the Zeeman interaction:

$$\mathcal{H}_{\mathrm{Z}} = -\boldsymbol{\mu} \cdot \boldsymbol{H}_0 = -\gamma\hbar\boldsymbol{I} \cdot \boldsymbol{H}_0 \tag{3.1}$$

This Hamiltonian can be diagonalized analytically, yielding eigenvalues $E_m = -\gamma\hbar H_0 m$ and corresponding eigenvectors $|m\rangle$. Here, $|m\rangle$ is the eigenstate of I_z with $m = -I, -I+1, \ldots, I$. The $2I+1$ energy levels are separated by equal intervals $\hbar\omega = \hbar\gamma H_0$, as illustrated in Fig. 3.1. When a radio-frequency (RF) field is applied perpendicular to $\boldsymbol{H}_0$ and its angular frequency matches ω, transitions occur between all the $|m\rangle \rightarrow |m+1\rangle$ states. This phenomenon is known as "**Nuclear Magnetic Resonance (NMR)**".

3.1.2 *Nuclear Magnetization*

In practice, we observe a many-body system of nuclear spins and the sum of nuclear magnetic moments (nuclear magnetization $\boldsymbol{M}$). The nuclear magnetization is expressed as:

K. Kinjo, *Nuclear Magnetic Resonance Study on Multiple Superconducting Phases in UTe2*, Springer Theses, https://doi.org/10.1007/978-981-97-8646-6_3

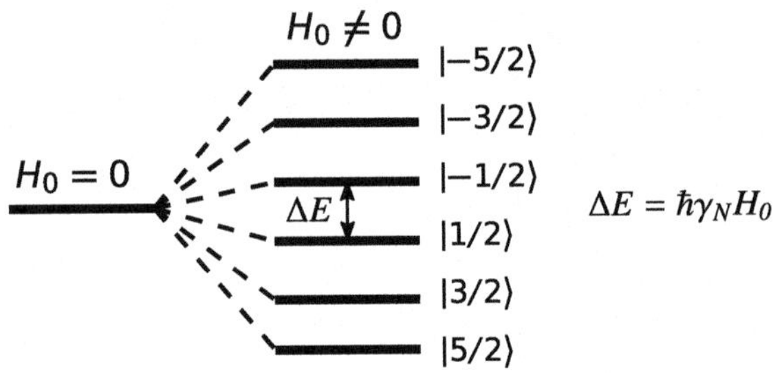

Fig. 3.1 Schematic image of the Zeeman split under magnetic field

$$\boldsymbol{M} = \gamma_n \hbar \sum_i \langle \boldsymbol{I}_i \rangle. \tag{3.2}$$

Here, $\boldsymbol{I}_i$ represents the nuclear spin at site i, and $\langle \ldots \rangle$ denotes the expectation value. In thermal equilibrium, the nuclear magnetization aligns with the z-axis such that $\boldsymbol{M} = M_0\hat{z}$, where $\hat{z}$ is the unit vector along the z-axis. The motion of an isolated nuclear spin is treated classically. The Heisenberg equation of motion for the Hamiltonian is:

$$\mathrm{i}\hbar \frac{\mathrm{d}I_i}{\mathrm{d}t} = [I_i, \mathcal{H}Z] = -\mathrm{i}\hbar\gamma \sum j, k\epsilon_{ijk} H_j I_k, \tag{3.3}$$

The expectation values of nuclear spin $\langle \boldsymbol{I} \rangle$ follow:

$$\frac{\mathrm{d}\langle \boldsymbol{I} \rangle}{\mathrm{d}t} = \gamma \langle \boldsymbol{I} \rangle \times \boldsymbol{H}. \tag{3.4}$$

This linear equation system is easily solved in a static field. In this scenario, the spin exhibits Larmor precession around the field direction (Fig. 3.2). The specific solutions with $\boldsymbol{H} = H_0\hat{z}$ are:

$$\langle I_x \rangle(t) = \langle I_x \rangle(0) \cos \omega_0 t + \langle I_y \rangle(0) \sin \omega_0 t, \tag{3.5}$$

$$\langle I_y \rangle(t) = \langle I_x \rangle(0) \sin \omega_0 t - \langle I_y \rangle(0) \cos \omega_0 t, \tag{3.6}$$

$$\langle I_z \rangle(t) = \langle I_z \rangle(0), \tag{3.7}$$

where $\omega_0 = \gamma H_0$ represents its angular frequency, which is equivalent to the energy-level interval.

The many-body nuclear spin system exhibits a macroscopic moment when a field is applied, and the spin susceptibility follows the Curie law: $\chi = C/T$, where C is a constant, and T is the temperature of the system. It is important to note that this is a high-temperature approximation, but it is practically valid in most experimental circumstances in solid-state physics.

The resonance of the macroscopic system becomes visible when the Larmor precession of the moment induces an electromotive force on a coil outside the sample.

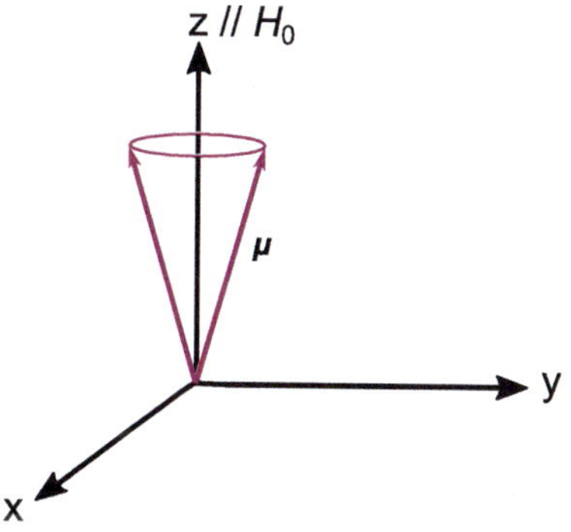

Fig. 3.2 Schematic image of the Larmor precession under the field

The macroscopic moment is aligned with the z direction at thermal equilibrium, but it rotates to the xy plane when the RF field is applied and continues to precess after the RF field is turned off. The signal intensity reaches its maximum when the RF field is applied in such a way that it rotates the moments by 90°. However, as shown later, it is challenging to utilize the 90° condition in actual experiments due to pulse heating.

3.1.3 *Free Induction Decay and Spin-Echo*

In real materials, the coherent motion of macroscopic nuclear magnetic moments decays within a finite time. This occurs because each nuclear spin undergoes a slightly different field compared to others, causing variations in the Larmor precession frequency. Causes of inhomogeneity include fields generated by electronic and nuclear spins, as well as inhomogeneity in the applied external field H_0. For example, if the field distribution is static and has a Lorentzian form, the NMR intensity (the electromotive force induced in the coil) decays exponentially as $V(t) \propto \cos(\omega_0 t) \exp(-t/T_2^*)$. This phenomenon is referred to as free induction decay (FID). Due to the finite dead time of the NMR spectrometer following the RF pulse field, measuring the FID signal can be challenging when T_2^* is extremely short. This issue might arise in correlated metals or magnetic materials.

The spin-echo technique is a common method for observing resonance signals in pulse NMR and effectively restores the coherence of nuclear moments. The motion of nuclear moments is shown in Fig. 3.3. In this approach, a second pulse RF field is applied after a time interval τ from the first RF pulse to cause a 180° rotation of the nuclear moments. Consequently, the signal is detected after a time of 2τ from the first $(\pi/2)$ pulse. This occurs because nuclear moments with positive phases are converted to negative phases, and vice versa. The time-dependent distribution of phases in the xy plane reaches a minimum at $t = 2\tau$. The spin-echo method allows

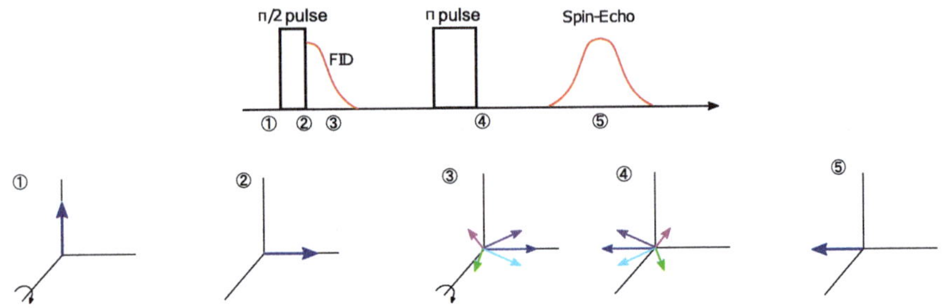

Fig. 3.3 Schematic image of the motion of the nuclear spins in NMR pulse sequence. First, the nuclear spins are excited by the applied $\pi/2$ pulse. **1–3** This coherence is abruptly destroyed in most cases of solid-state measurements. FID signal is observed here. To observe a spin-echo signal, π pulse is needed. After the π pulse, the motion of the nuclear spin flipped and again become coherent (**4–5**)

for avoiding dead time and observing the NMR signal with an efficient time delay from the (second) RF pulse.

It is important to note that the signal intensity relies on the time interval between pulses τ due to the time-dependent inhomogeneity (fluctuation) of the internal field and the presence of nuclear spin interactions. The decay time of the spin echo is denoted as T_2, and $1/T_2$ is referred to as the "spin-echo decay rate," "nuclear spin-spin relaxation rate," or "transverse relaxation rate." The value $1/T_2$ represents a more intrinsic width of the field than $1/T_2^*$.

3.2 Application for Solid-State Physics

NMR is the technique for studying the electronic state of materials from the microscopic point of view. The nuclei couple to the electrons through a hyperfine interaction. In this section, I will explain the NMR quantities for studying solid-state physics.

3.2.1 Hyperfine Interaction

The nuclei couple to the electrons through a hyperfine interaction. We get the microscopic information of the electronic state from the nuclei through the hyperfine interaction.

We derive the Hamiltonian for hyperfine interaction as follows. By expanding the vector potential arising from the nuclear current, we obtain

$$
\begin{aligned}
\boldsymbol{A}(\boldsymbol{r}) &= \int d\boldsymbol{r}_{\mathrm{n}} \frac{\rho(\boldsymbol{r}_{\mathrm{n}})\boldsymbol{v}(\boldsymbol{r}_{\mathrm{n}})}{c|\boldsymbol{r}-\boldsymbol{r}_{\mathrm{n}}|} \\
&= \frac{1}{cr}\int d\boldsymbol{r}_{\mathrm{n}}\rho(\boldsymbol{r}_{\mathrm{n}})\boldsymbol{v}(\boldsymbol{r}_{\mathrm{n}}) \\
&+ \frac{1}{2cr^3}\int d\boldsymbol{r}_{\mathrm{n}} \times [(\boldsymbol{r}\cdot\boldsymbol{r}_{\mathrm{n}})\boldsymbol{v}(\boldsymbol{r}_{\mathrm{n}}) + \{\boldsymbol{r}\cdot\boldsymbol{v}(\boldsymbol{r}_{\mathrm{n}})\}\boldsymbol{v}(\boldsymbol{r}_{\mathrm{n}})] \\
&- \frac{\boldsymbol{r}}{r^3} \times \frac{1}{2c}\int d\boldsymbol{r}_{\mathrm{n}}\rho(\boldsymbol{r}_{\mathrm{n}})\{\boldsymbol{r}_{\mathrm{n}} \times \boldsymbol{v}_{\mathrm{n}}\} \\
&+ \cdots .
\end{aligned}
\tag{3.8}
$$

When the current distribution is steady with the axis of angular momentum, the first two terms disappear and only the third term remains. The integral of the third term is nuclear dipole moment $\boldsymbol{\mu}_I$ and $\boldsymbol{\mu}_{\mathrm{I}}$ is written as

$$
\boldsymbol{\mu}_I = g_{\mathrm{N}}\mu_{\mathrm{N}}\boldsymbol{I} = \gamma_{\mathrm{N}}\hbar\boldsymbol{I}. \tag{3.9}
$$

Then, the vector potential is written as

$$
\boldsymbol{A}(\boldsymbol{r}) = \boldsymbol{\mu}_I \times \frac{\boldsymbol{r}}{r^3}. \tag{3.10}
$$

Substituting this into the Hamiltonian, we get

$$
\mathcal{H}_{\mathrm{hf}} = \frac{\boldsymbol{p}^2}{2m} - \frac{3}{mc}\frac{(\boldsymbol{\mu}_I \times \boldsymbol{r})\cdot\boldsymbol{p}}{r^3} + \frac{e\hbar}{2mc}\boldsymbol{S}\cdot\left\{\frac{\boldsymbol{\mu}_I}{r^3} - 3\frac{(\boldsymbol{r}\cdot\boldsymbol{\mu}_I)\boldsymbol{r}}{r^5}\right\} \tag{3.11}
$$

The first term gives constant energy shift and this is ignorable. A modification of the second term is

$$
2\mu_{\mathrm{B}}\frac{\boldsymbol{\mu}_I\cdot\boldsymbol{l}}{r^3}. \tag{3.12}
$$

This term is called orbital hyperfine interaction. The third term is the dipolar hyperfine interaction. The contribution of the s-electrons to these hyperfine interactions is apparently zero. On the other hand, since Eq. 3.8 neglected the charge distribution at the nuclear position, there is an s-electron contribution that is not included in Eq. 3.8. Consider the matrix elements of hyperfine interaction for $\psi(\boldsymbol{r})$, the matrix elements are written as

$$
\begin{aligned}
&-\frac{e\hbar}{2mc}\int d\boldsymbol{r}\psi^*(\boldsymbol{r})\boldsymbol{S}\cdot\nabla\times\boldsymbol{A}(\boldsymbol{r})\psi(\boldsymbol{r}) \\
&= -\frac{e\hbar}{2mc}\int_{r<R} d\boldsymbol{r}\psi^*(\boldsymbol{r})\boldsymbol{S}\cdot\nabla\times\boldsymbol{A}(\boldsymbol{r})\psi(\boldsymbol{r}) \\
&\quad -\frac{e\hbar}{2mc}\int_{r>R} d\boldsymbol{r}\psi^*(\boldsymbol{r})\boldsymbol{S}\cdot\nabla\times\boldsymbol{A}(\boldsymbol{r})\psi(\boldsymbol{r}),
\end{aligned}
\tag{3.13}
$$

where R is the radius for nuclei. The second term is the same as the Eq. 3.11, then ignorable. By transforming the first term, we get

$$
\begin{aligned}
-\frac{e\hbar}{2mc}\int_{r<R} d\boldsymbol{r}\psi^*(\boldsymbol{r})\boldsymbol{S}\cdot\nabla\times\boldsymbol{A}(\boldsymbol{r})\psi(\boldsymbol{r}) &= -\frac{e\hbar}{2mc}\int_{r<R} d\boldsymbol{r}\nabla\cdot(\boldsymbol{A}\times\boldsymbol{S})|\psi(\boldsymbol{r})|^2 \\
&= -\frac{e\hbar}{2mc}\boldsymbol{S}\cdot\int d\boldsymbol{n}\times\boldsymbol{A}|\psi(\boldsymbol{r})|^2 \\
&= -\frac{e\hbar}{2mc}\boldsymbol{S}\cdot\int\frac{\boldsymbol{R}\times(\boldsymbol{\mu_I}\times\boldsymbol{R})}{R^2}d\Omega|\psi(0)|^2 \\
&= -\frac{8\pi}{3}\frac{e\hbar}{2mc}\boldsymbol{S}\cdot\boldsymbol{\mu_I}|\psi(0)|^2
\end{aligned}
\tag{3.14}
$$

This term is known as Fermi contact. Summarize these interactions, we get the simple hyperfine Hamiltonian as follows:

$$
\mathcal{H}_{\mathrm{hf}} = -\gamma_{\mathrm{n}}\hbar\boldsymbol{I}\cdot\boldsymbol{H}_{\mathrm{hf}}, \tag{3.15}
$$

where $\boldsymbol{H}_{\mathrm{hf}}$ is so-called hyperfine field and written as

$$
\boldsymbol{H}_{\mathrm{hf}} = \gamma_{\mathrm{e}}\hbar\sum_i\frac{\mu_0}{4\pi}\left\{-\frac{8\pi}{3}\boldsymbol{S}_i\delta(\boldsymbol{r}_i)+\frac{1}{r_i^3}\left(\boldsymbol{S}_i-\boldsymbol{L}_i-\frac{3(\boldsymbol{S}_i\cdot\boldsymbol{r}_i)\boldsymbol{r}_i}{r_i^2}\right)\right\} \tag{3.16}
$$

In many cases, the Fermi contact term and core polarization term are more effective than the magnetic dipole interaction, so the spin part is considered to be

$$
-\gamma_{\mathrm{n}}\hbar H_{\mathrm{hf},i} = \sum_j A_{ij}\boldsymbol{S}_j, \tag{3.17}
$$

or

$$
\mathcal{H}_{\mathrm{hf}} = \sum_{ij} I_i A_{ij}\boldsymbol{S}_j. \tag{3.18}
$$

A is a second-order tensor.

3.2.2 Knight Shift K

As mentioned above, the hyperfine interaction is treated as the additional field:

$$
\mathcal{H}_{\mathrm{Z}} + \mathcal{H}_{\mathrm{hf}} = -\gamma_{\mathrm{n}}\hbar\boldsymbol{I}\cdot(\boldsymbol{H}_0+\boldsymbol{H}_{\mathrm{hf}}). \tag{3.19}
$$

In general, the motion of the electron system is much faster than the dynamics of the nuclear spins, so a time average of this extra magnetic field is observed. This causes the NMR resonance lines to shift from the position expected from the external magnetic field. In other words, the resonance frequency f_0 shifts to f_{res}. The value of this shift, normalized by f_0, is called the Knight shift:

$$K \equiv \frac{f_{\mathrm{res}} - f_0}{f_0}. \tag{3.20}$$

The Knight shift has a strong connection with the electronic spin susceptibility, denoted as χ_{spin}, and can be expressed in the form:

$$K = A_{\mathrm{hf}}^{\mathrm{spin}} \chi_{\mathrm{spin}} + K_{\mathrm{orb}}. \tag{3.21}$$

In this context, it's crucial to factor in the hyperfine coupling constant, A_{hf}, and the orbital shift, K_{orb}, which serves a purpose similar to the chemical shift observed in molecular structures. While the above formula is applicable for systems with a single orbital, the dynamics are considerably more intricate for systems encompassing multiple orbitals. The nature of the coupling can cause A_{hf} to reflect a specific local symmetry at the nuclear site, leading to significant anisotropy.

In the simplest case, electrons show Pauli paramagnetism, which is temperature-independent and results in a constant Knight shift. However, in reality, the electronic spin susceptibility demonstrates temperature dependence, allowing us to experimentally determine the hyperfine coupling constant by comparing K and χ with temperature as an implicit parameter. This technique is referred to as the K–χ plot.

As mentioned, the Knight shift reflects the bulk susceptibility, but it presents distinct meanings, offering additional insights. The Knight shift gives a local magnetic susceptibility. That is, we can observe the local magnetic field in the magnetically ordered state and the spin susceptibility in the superconducting state as discussed later.

3.2.3 *Nuclear Spin-Lattice Relaxation Rate* $1/T_1$

In the previous section, the Knight shift provided static information about the electronic state; however, NMR can also measure quantities that exhibit dynamic information, which is intrinsic to the electronic system. One important measured quantity is the nuclear spin-lattice relaxation time, denoted as T_1, or inversely, the nuclear spin-lattice relaxation rate, expressed as $1/T_1$.

The hyperfine interactions implicate both the phenomena of nuclear relaxation and the Knight shift. Upon excitation by a radiofrequency (RF) field, the nuclear ensemble is destroyed from thermal equilibrium, becoming a state of non-equilibrium. This deviation is transitory, marked by the nuclear system's interaction with the electron system, functioning as a thermal bath. Consequently, a restitution of thermal equilibrium ensues, characterized by a temporal duration delineated as T_1. This process represents the exchange of energy between the disturbed nuclear spins and the surrounding electronic environment, highlighting the dynamic relationship within the system.

From microscopic view, nuclear relaxation is conceptualized as a scattering process where nuclear and electronic spins mutually flip each other. In standard metals (such as copper), electrons possessing Fermi energy within roughly $k_{\mathrm{B}}T$ are engaged in this scattering, leading to a $1/T_1$ ratio that aligns proportionally with T in straightforward metallic contexts. More generally, a theoretical relationship for the magnetic origin of $1/T_1$ is derived from linear response theory, and the outcome is as follows:

$$\begin{aligned}\frac{1}{T_1} &= \frac{1}{2}\sum_{\boldsymbol{q}}(A_{\boldsymbol{q}}A_{-\boldsymbol{q}})_{\perp}\int_{-\infty}^{\infty}\langle\delta S_{+}(-\boldsymbol{q},t)\delta S_{-}(\boldsymbol{q},0)\rangle\,\mathrm{e}^{-\mathrm{i}\omega_{\mathrm{n}}t}\mathrm{d}t \\ &= \frac{2k_{\mathrm{B}}T}{\hbar(2\mu_{\mathrm{B}})^2}\sum_{\boldsymbol{q}}(A_{\boldsymbol{q}}A_{-\boldsymbol{q}})_{\perp}\frac{\chi_{\perp}''(\omega_{\mathrm{n}},\boldsymbol{q})}{\omega_{\mathrm{n}}}. \end{aligned} \tag{3.22}$$

Here, $\chi_{\perp}''(\omega,\boldsymbol{q})$ is an imaginary part of the dynamic electronic spin susceptibility perpendicular to the magnetic field, $A_{\boldsymbol{q}}$ is a momentum-dependent hyperfine coupling constant ($\boldsymbol{q}=0$ component corresponds to the $\boldsymbol{q}$-independent one), and ω_{n} is an angular frequency of the NMR line. Because $\chi_{\perp}''(\omega,\boldsymbol{q})$ is a quantity calculated by various theoretical models, this formula directly connects to the theoretical explanations and experimental results. Some important features of $1/T_1$ are:

- $1/T_1$ is sensitive to fluctuations occurring at $\omega=\omega_{\mathrm{n}}$, typically situated at the lower end of the energy spectrum in electronic systems.
- It reflects fluctuations that are perpendicular to the applied magnetic field (or, more precise, perpendicular to the quantization axis of nuclear spin).
- Essentially, $1/T_1$ is responsive to all wave vector components, although without the capability to individually resolve these vectors.

For comparison between Knight shift and $1/T_1$, K is proportional to the $\chi(\boldsymbol{q}=0,\omega=0)$ and $1/T_1$ is proportional to the $\sum_{\boldsymbol{q}}\chi(\boldsymbol{q},\omega)$.

3.3 NMR in Superconductors

3.3.1 NMR Knight Shift

Spin-Singlet Case

Form Sect. 3.2.2, Knight shift reflects the spin susceptibility. In the spin-singlet superconductors, thermally excited quasiparticles only affect the spin susceptibility since spin-singlet Cooper pairs are $S=0$ state. Consider the response to the magnetic field in the BCS superconductors. The partition function is written as

$$Z_k = 1 + e^{-\beta(E_k - \mu_B H)} + e^{-\beta(E_k + \mu_B H)} + e^{-2\beta E_k} \tag{3.23}$$

$$\simeq \left(1 + e^{-\beta E_k}\right)^2. \tag{3.24}$$

Then, magnetization M is calculated

$$\begin{aligned} M &= \sum_k \frac{\mu_B e^{-\beta(E_k - \mu_B H)} - \mu_B e^{-\beta(E_k + \mu_B H)}}{\left(1 + e^{-\beta E_k}\right)^2} \\ &\simeq 2{\mu_B}^2 H \sum_k \beta \frac{e^{-\beta E_k}}{\left(1 + e^{-\beta E_k}\right)^2} \\ &= {\mu_B}^2 \sum_k \frac{\beta}{2\cosh^2 \frac{\beta E_k}{2}} H \\ &= \mu_B^2 H N(0) \int_0^\infty d\xi \frac{\beta}{2\cosh^2 \frac{\beta E_k}{2}}, \quad E_k \equiv \sqrt{\xi^2 + \Delta^2}. \end{aligned} \tag{3.25}$$

This reproduces the spin susceptibility of the normal state in the limit of $\Delta \to 0$. That is,

$$\chi_n = \mu_B^2 N(0) \int_0^\infty d\xi \frac{\beta}{2\cosh^2 \frac{\beta\xi}{2}} = 2\mu_B^2 N(0). \tag{3.26}$$

Then, the temperature dependence of the spin susceptibility is written as follows using normal-state susceptibility and Yosida function ($Y(T)$),

$$\chi(T) = \chi_n \int_0^\infty d\xi \frac{1}{2k_B T \cosh^2 \frac{E}{2k_B T}} \equiv \chi_n Y(T). \tag{3.27}$$

Figure 3.4a shows the Yosida function.

Spin-Triplet Case

For spin-triplet Cooper pairs, Cooper pairs have spin $S = 1$, then spin susceptibility remains constant below T_c without spin-orbit coupling. With spin-orbit coupling, the order parameter is written in d-vector, which is defined in Chapter 1. Then spin susceptibility decreases when the d-vector was locked in the crystal and the field is parallel to the d-vector (perpendicular to the spin). A schematic image for the spin susceptibility of spin-triplet superconductors is shown in Fig. 3.4b.

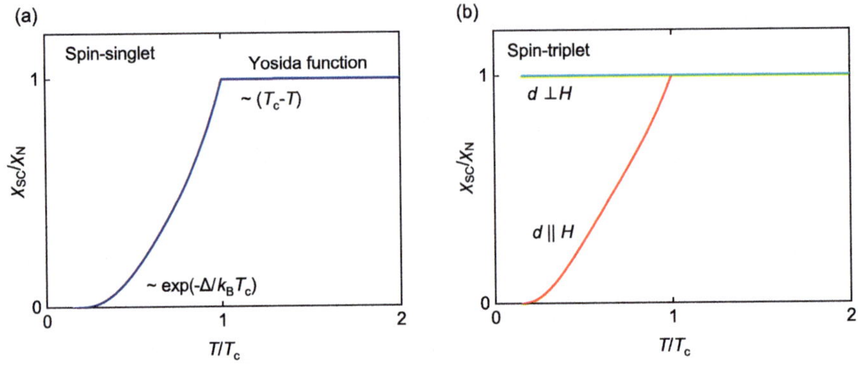

Fig. 3.4 **a** For spin-singlet case. The shape of the Yosida function. Near T_c, spin susceptibility decreases linearly, and exponentially decays near 0 K, reflecting the gap structure. **b** For spin-triplet case. When the field is perpendicular to the d-vector, spin susceptibility does not change, on the other hand, when the field is parallel to the d-vector, spin susceptibility decreases

3.3.2 *Nuclear Spin-Lattice Relaxation Rate*

As mentioned in Sect. 2.2.3, $1/T_1T$ is a physical quantity that can detect DOS. In SC states, $1/T_1T$ can detect DOS of excited quasiparticles. $1/T_1T$ in the SC state can be expressed by the following equation.

$$1/T_1T = \frac{\pi A_{\text{hf}}^2}{\hbar N^2}\int_0^\infty \left(1+\frac{\Delta^2}{E^2}\right)N_s^2(E)f(E)(1-f(E))dE \tag{3.28}$$

Here, $N_s^2(E)$ is quasiparticle DOS. Figure 3.5 shows the temperature dependence of the $1/T_1T$ in the SC state with three different nodal structures (full gap, line node, and point node). Here, $1/T_1T$ is normalized by the value at T_c.

The factor $\left(1+\frac{\Delta^2}{E^2}\right)$ in the Eq. 3.28 is called "coherence factor". The coherence factor is greater than 1 when there is no significant change in the gap (s-wave superconductor with a full gap). As a result, a peak occurs near T_c as shown in Fig. 3.5. The peak is called "coherence peak" and is an indication of conventional isotropic s-wave superconductors. On the other hand, in the case of unconventional superconductors with an anisotropic SC gap, the coherence factor becomes 1 due to the sign change of the SC gap, which results in no (or tiny) coherence peak occurring. The temperature dependence at low temperatures depends on the nodal structure as follows.

$$1/T_1T \propto \begin{cases} \exp\left(-\frac{\Delta}{k_BT}\right) & \text{(full gap)} \\ T^3 & \text{(line node)} \\ T^5 & \text{(point node)} \end{cases}$$

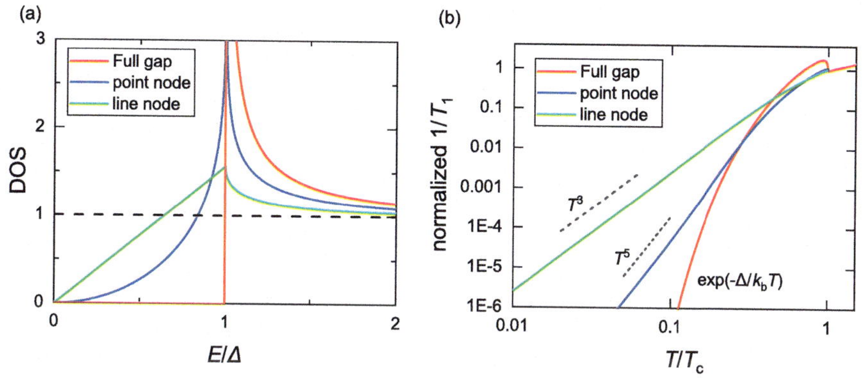

Fig. 3.5 **a** The energy dependence of the quasiparticle DOS of each superconducting gap structure. The full gap superconductivity shows strong divergence at $E = \Delta$. **b** Calculated $1/T_1$ using quasiparticle DOS in (**a**)

Therefore, $1/T_1T$ is a good quantity to investigate the SC symmetry and the nodal structure.

Chapter 4
Experimental

4.1 Samples

We used two single-crystal samples #1 and #2 provided by D. Aoki group in IMR, Tohoku University. These samples were grown using the chemical transport method with iodine as the transport agent. Both samples are 99.9% ^{125}Te enriched. Sample #1 was used for the pressure study, and #2 was used for the high-field study. The sample #1 is a rectangular shape and the size is $2.0 \times 1.0 \times 0.5$ mm^3. The sample #2 is a stacked hexagonal shape and the size is $5.0 \times 3.0 \times 1.0$ mm^3 (Fig. 4.1).

These samples are characterized by the specific heat measurements and χ_{AC} measurements, as shown in Fig. 4.2. While a sharp specific heat jump is visible, the residual specific heat term is about 50% for sample #2, and the transition temperature is about 1.6 K for both samples, which is lower than the ideal 2.1 K, suggesting a slight U-vacancy as previously reported [1].

4.2 NMR Experiment

The frequency-swept NMR spectrum was obtained using the Fourier transform of a spin-echo signal observed after the spin-echo RF pulse sequence in a fixed magnetic field. Because of the presence of the two inequivalent crystallographic Te sites in UTe_2, we observed two Te signals in each magnetic field direction, as shown in Fig. 4.3. We call the one with a large [small] K Te2 [Te1] as defined in the first single-crystal NMR reports [2]. The magnetic field was calibrated using the ^{65}Cu($^{65}\gamma_n/2\pi = 12.089$MHz/T) and ^{63}Cu $\left(^{63}\gamma_n/2\pi = 11.285\ \text{MHz/T}\right)$NMR signals from the NMR coil. The NMR spectrum in the SC state was recorded with the field-cooling process. ^{125}Te-nuclear spin-lattice relaxation time T_1 was determined by fitting the time dependence of the spin-echo intensity after the saturation of the nuclear magnetization M. The fitting function is a simple exponential function in the

K. Kinjo, *Nuclear Magnetic Resonance Study on Multiple Superconducting Phases in UTe2*, Springer Theses, https://doi.org/10.1007/978-981-97-8646-6_4

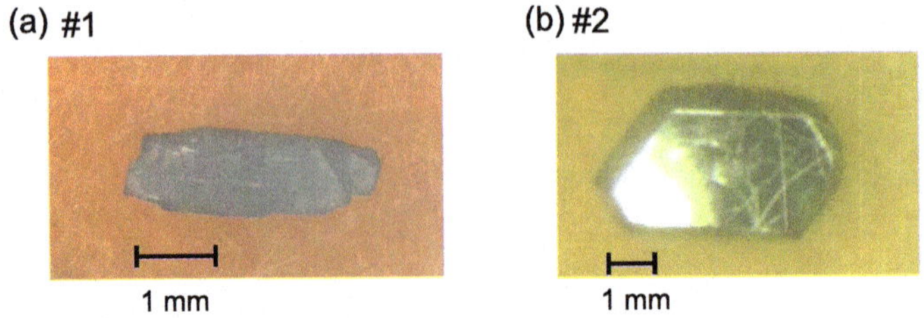

Fig. 4.1 Photographs of UTe_2 samples used in this study

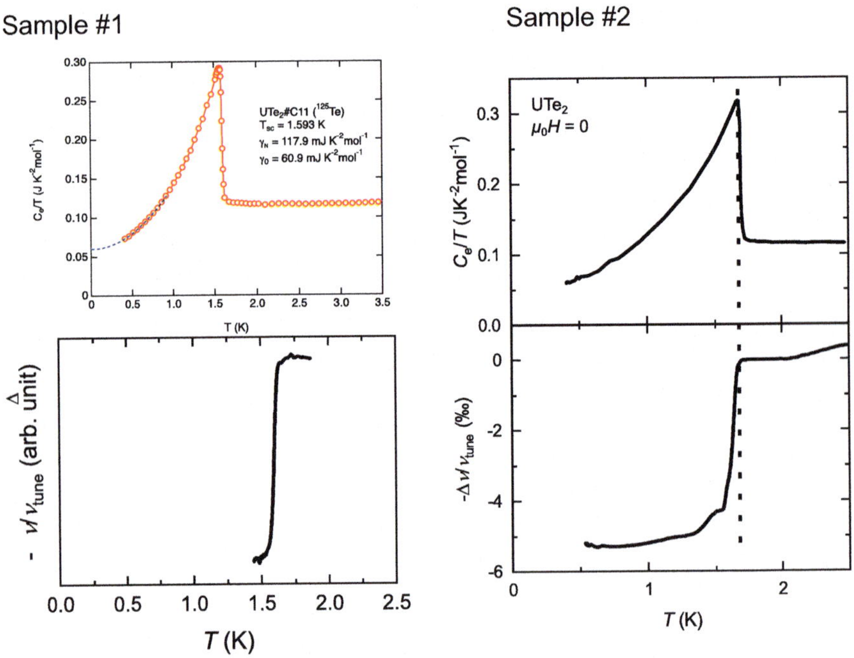

Fig. 4.2 The temperature dependence of specific heat measurements and χ_{AC}. The transition temperature of the sample #1 is 1.6 K and of the sample #2 is 1.65 K

case of $I = 1/2$, and the typical relaxation curve $R(t)$ of the nuclear magnetization is defined as

$$R(t) \equiv [M(\infty) - M(t)]/M(\infty) \tag{4.1}$$

at time t after the saturation of M. Low-temperature NMR measurements down to 150 mK for sample #1 were performed using a ^{3}He–^{4}He dilution refrigerator, and the pressure cell was immersed into the mixture to reduce the heating effect in the low-temperature measurements. For sample #2, measurements were performed using a ^{3}He refrigerator and the sample was also immersed into the ^{3}He liquid.

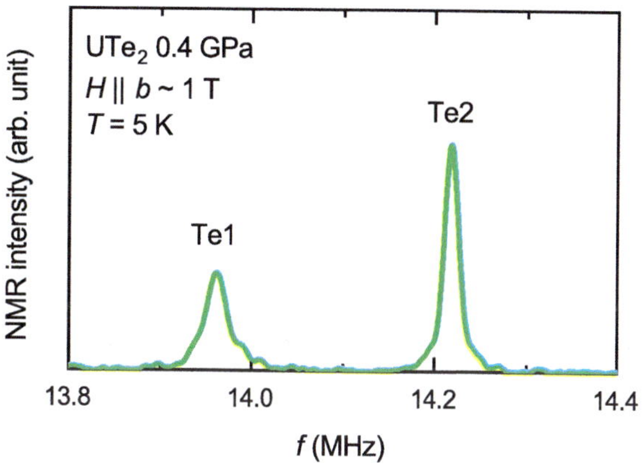

Fig. 4.3 Typical NMR spectrum measured for $H \parallel b$. The definition of Te1 and Te2 is discussed in the main text

4.3 Heat-Up Effect by NMR RF Pulse

Although the energy of RF pulse of NMR measurements as electromagnetic waves is low enough to avoid changing the electronic state, the heat-up of the sample due to H1 is a problem that cannot be ignored. It is known that the heat-up caused by NMR pulses returns to thermal equilibrium in the order of microseconds, and is difficult to detect by ordinary methods. Here, we used the phase detection pulse method as described in Ref. [3] to confirm that the measured spectrum shows the superconducting state. In this method, a long, weak pulse is input at the position where the NMR spectrum should be observed (detection pulse in Fig. 4.4). Even without the pulse, a change in the temperature dependence is visible due to the change in impedance caused by the SC transition. Therefore, we can add an NMR pulse sequence and show that the measurement is performed in an SC state by not moving from the position where there is no RF pulse and is indeed in superconductivity. It is important to note that if the change is in the vicinity of -15 dB in this case, it is superconducting. Still, we cannot rule out the possibility that it has heated up and is just below the transition temperature. In this case, a pulse of 9.5 μJ was used, and it is clear from the Fig. 4.4 that a sufficiently weak pulse was used.

Previously, this measurement technique had only been used at ambient pressure and with the sample in good contact with the heat bath. This time, the measurement can be performed in a pressure cell, where the cooling power is low. Even in such a situation, low-RF pulse power assures superconductivity.

Comparison with Previous Measurements

The heat-up effect was also checked on the NMR measurements at ambient pressure. Figure 4.5a and b show the RF-pulse energy dependence of the output of the phase detection pulse at ambient pressure under the b-axis and a-axis fields, respectively. As shown in the figure, there is a clear difference between superconducting and non-superconducting states. In this measurement, the RF-pulse energy depends on the tuning of the NMR tank circuit on a power amplifier, so the value of the

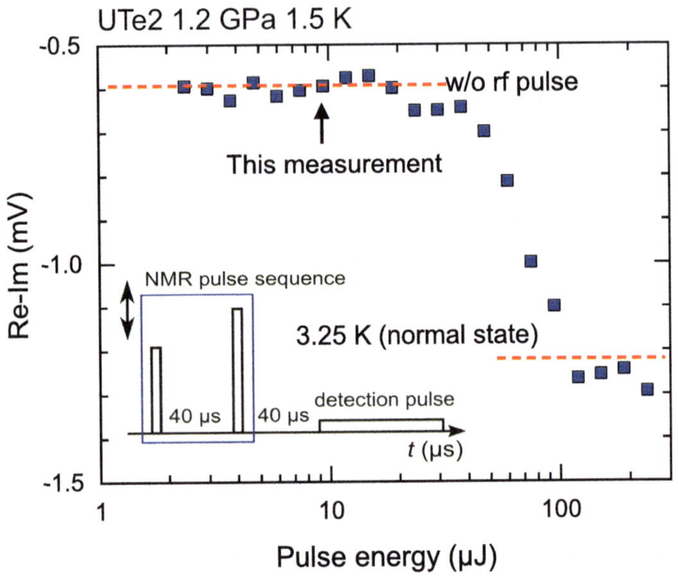

Fig. 4.4 The pulse energy dependence of the output voltage of phase detection pulses. The measurements were performed in the SC state at 1.5 K and at 1.0 T under 1.2 GPa. In this measurement, NMR RF pulses were applied at a power of 9.5 μJ (shown by arrows) to not affect superconductivity. The inset shows a schematic of the NMR pulse sequence. The heating of the sample is confirmed by varying the voltage of the two pulses enclosed in the blue box. These measurements were done in the pressure cell

RF-pulse energy is highly dependent on an experimental spectrometer. In addition, Fig. 4.5a and b represent the same results, although the direction of change is opposite. According to the papers, pulses with a magnitude of about 100 μJ in the b- and c-axis measurements and 344 μJ in the a-axis direction are used [4, 5]. In the a-axis direction, the pulse energy is relatively large. This is due to the fact that it is very difficult to obtain a signal in the a-axis direction since $1/T_2$, the nuclear spin-spin relaxation rate, diverges at low temperatures for $H \parallel a$. In any case, it is proven that NMR measurement was done in the superconducting state.

For the measurements of the high field heat-up check (Chap. 7), a similar experiment could not be performed due to the lack of measurement equipment in the high-field facility of IMR at Tohoku University. However, we indirectly confirmed that reliable measurements were performed from the dependence on the NMR pulse power and the temperature dependence of the spectrum.

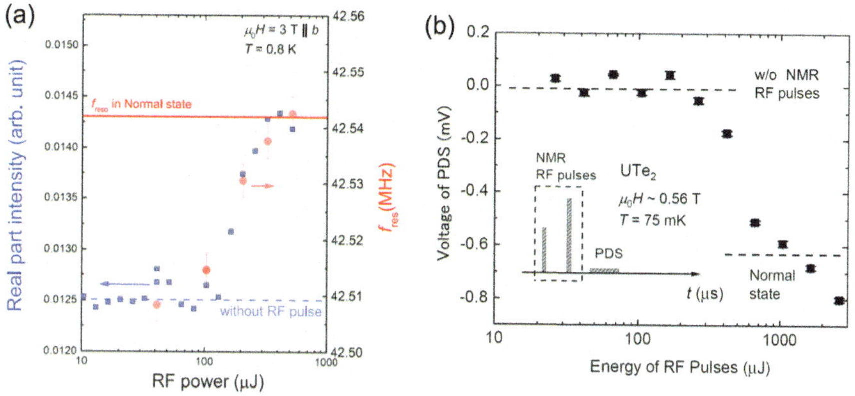

Fig. 4.5 The pulse energy dependence of the output voltage of phase detection pulses at ambient pressure for **a** $H \parallel b$ and **b** $H \parallel a$. The NMR measurements for $H \parallel b$ and c were done with a pulse power of about 50 μJ, and for $H \parallel a$ were done with a pulse power of about 200 μJ. These figures are reproduced with CC-BY 4.0 license from Ref. [4] © (2019) and [5]

4.4 AC Susceptibility Measurements

The AC susceptibility was measured by recording the resonance frequency of the NMR tank circuit during cooling. Since the self-inductance of the NMR coil L is proportional to the permeability, the change of the resonant frequency $\Delta f = (\Delta LC)^{-1}$ reflects the change of the susceptibility. The frequencies used in the studies are 12–14 MHz.

4.5 Hydrostatic Pressure

Hydrostatic pressure was applied using a piston-cylinder-type cell [6]. The outer cylinder was made from CuBe-NiCrAl alloy and the inner piston was made from tungsten carbide. The sample and the NMR coil were inside a Teflon tube and the plug was sealed by Stycast 2850FT. The inside of the tube was filled with Daphne 7373 (oil) as a pressure mediate (Fig. 4.6).

The applied pressure is estimated using the T_c of Pb metal. T_c of Pb linearly decreases with increasing pressure up to 5 GPa, and written as

$$P\,[\mathrm{GPa}] = \frac{T_c(0)\,[\mathrm{K}] - T_c(P)\,[\mathrm{K}]}{0.364\,[\mathrm{K}\cdot\mathrm{GPa}^{-1}]}, \tag{4.2}$$

where $T_c(0) = 7.181$ K.

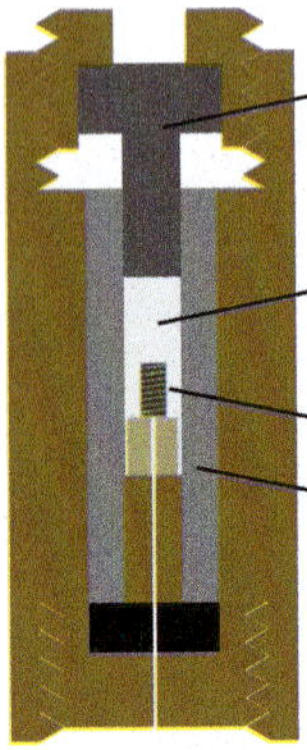

Fig. 4.6 Schematic image of pressure cell

References

1. Haga Y, Opletal P, Tokiwa Y, Yamamoto E, Tokunaga Y, Kambe S, Sakai H (2022) Effect of uranium deficiency on normal and superconducting properties in unconventional superconductor UTe2. J Phys: Conden Matter 34:175601
2. Tokunaga Y, Sakai H, Kambe S, Hattori T, Higa N, Nakamine G, Kitagawa S, Ishida K, Nakamura A, Shimizu Y, Homma Y, Li D, Honda F, Aoki D (2019) 125Te NMR study on a single crystal of heavy fermion superconductor UTe2. J Phys Soc Jpn 88:073701
3. Kinjo K, Manago M, Kitagawa S, Mao ZQ, Yonezawa S, Maeno Y, Ishida K (2022) Superconducting spin smecticity evidencing the Fulde-Ferrell-Larkin-Ovchinnikov state in Sr2RuO4. Science 376:397–400
4. Nakamine G, Kitagawa S, Ishida K, Tokunaga Y, Sakai H, Kambe S, Nakamura A, Shimizu Y, Homma Y, Li D, Honda F, Aoki D (2019) Superconducting properties of heavy Fermion UTe2 revealed by 125Te-nuclear magnetic resonance. J Phys Soc Jpn 88:113703
5. Fujibayashi H, Nakamine G, Kinjo K, Kitagawa S, Ishida K, Tokunaga Y, Sakai H, Kambe S, Nakamura A, Shimizu Y, Homma Y, Li D, Honda F, Aoki D (2022) Superconducting order parameter in UTe2 determined by knight shift measurement. J Phys Soc Jpn 91:043705
6. Fujiwara N, Môri N, Uwatoko Y, Matsumoto T, Motoyama N, Uchida S (2003) Superconductivity of the Sr2Ca12Cu24O41 spin-ladder system: are the supercon ducting pairing and the spin-gap formation of the same origin? Phys Rev Lett 90:137001

Chapter 5
Results

5.1 SC Order Parameter in UTe_2 Determined by Knight Shift Measurement

Short Summary

In this section, we performed the NMR measurements on UTe_2 to investigate the spin susceptibility in the SC state. A clear anomaly in the temperature dependence of the NMR Knight shift was found for $H \parallel b$ and c, whereas only a small anomaly was found for $H \parallel a$. The anomaly for $H \parallel b$ and c is much larger than the effect of superconducting diamagnetism, suggesting a distinct decrease in the NMR Knight shift. On the other hand, the a-axis Knight shift exhibits only a reduction in the amount that can be explained by superconducting diamagnetism, indicating that the change in spin susceptibility due to superconductivity is almost zero. These results indicate that superconducting spin is along the a axis in the LFSC state, consistent with normal state properties.

5.1.1 Sample

This section was based on the Refs. [1–3]. First of all, it should be mentioned that these measurements are based on first-generation samples, i.e., samples with transition temperatures around 1.6 K (Fig. 5.1). As we will see later in the discussion, the NMR results for the 1.6 and 2.1 K samples are different for $H \parallel a$ [4], and we must be careful about the results due to impurities.

K. Kinjo, *Nuclear Magnetic Resonance Study on Multiple Superconducting Phases in UTe2*, Springer Theses, https://doi.org/10.1007/978-981-97-8646-6_5

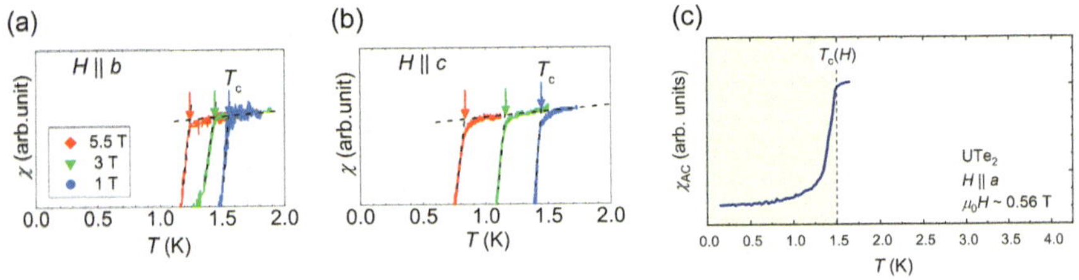

Fig. 5.1 Temperature dependence of AC-susceptibility using NMR coils. **a** for $H \parallel b$, **b** for $H \parallel c$, and **c** for $H \parallel a$

5.1.2 Spin Susceptibility in SC State

Figure 5.2 demonstrates the distinct responses in the NMR spectra, dependent on the orientation of the applied magnetic field, particularly noticeable around the critical temperature, T_c:

1. **a-axis spectrum:** Only a minute variation is observable around T_c. This minor change signals relatively stable magnetic properties in the system along this axis as it transitions through T_c.
2. **b-axis spectrum:** The Knight shift at the peak decreases. In addition, a pronounced broadening of the spectrum becomes evident below T_c. This characteristic widening suggests the emergence of magnetic anomalies, possibly due to the presence of minuscule impurities that slightly disturb the magnetic properties.

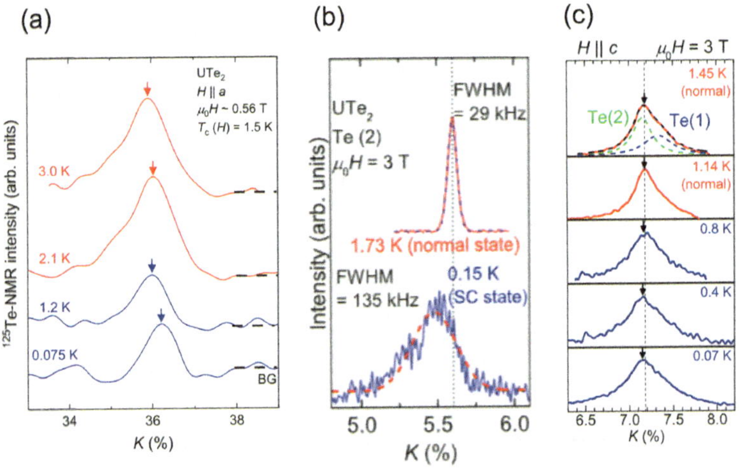

Fig. 5.2 The temperature variation of the NMR spectrum for **a** $H \parallel a$, **b** $H \parallel b$, and **c** $H \parallel c$. These figures are reproduced with permission from Ref. [1, 3] © (2019, 2022) The Physical Society of Japan, and Ref. [2] © (2021) American Physical Society

3. **c-axis spectrum:** There is a subtle shift in the NMR Knight shift observable around T_c, similar to the a-axis. However, a significant aspect of the c-axis spectrum is the substantial increase in the Full Width at Half Maximum (FWHM), attributed to the overlapping NMR signals from the Te1 and Te2 nuclei, which results in this noticeable broadening.

These spectral responses suggest the subtle changes in the local magnetic environments along different crystallographic directions, as manifested through their respective NMR spectra. The disparities become particularly insightful when examining the system's transition through the critical temperature, T_c.

As is clear from this figure, a comparison of the NMR spectra just above T_c and at the lowest temperature shows that the Knight shift clearly decreases in the b and c directions, while the Knight shift does not decrease in the a direction.

To enhance the clarity of the change in the NMR Knight shift, the diamagnetic component and the temperature dependence of the normal-state component are subtracted in Fig. 5.3. It is obtained as $\Delta K = K_S - K_N$. K_N is subtracted by a linear function (a, c) or as a quadratic function (b). As more clearly shown, the Knight shift decreases by about 0.05% in the b, c direction, whereas it does not decrease within the error bars in the a direction. Note that, even if we are using a very small magnetic field of 0.56 T, the limit of measurable field, we need to consider the possibility that d-vector rotation may be occurring.

Now, we discuss the change in the NMR Knight shift. In NMR studies of conventional materials, the spin part of the Knight shift has been evaluated from the K-χ plot. However, it was clarified that this simple estimation is not valid in heavy-fermion superconductors due to the strong spin-orbit coupling, particularly for the estimation of the spin-part K related to the superconductivity. In such a case, ΔK in the superconductivity can be estimated from the change of the Sommerfeld coefficient γ_e as follows:

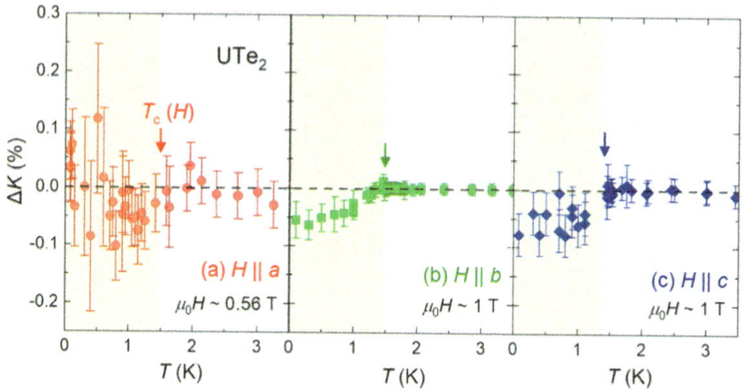

Fig. 5.3 The temperature dependence of ΔK_i $(i = a, b, c)$. Definition of ΔK is written in the main text

$$\Delta K_{\mathrm{spin}} = \frac{A_{\mathrm{hf}}}{N_{\mathrm{A}}\mu_{\mathrm{B}}}\Delta\chi^{\mathrm{qp}} = \frac{A_{\mathrm{hf}}}{N_{\mathrm{A}}\mu_{\mathrm{B}}}\frac{\Delta\gamma_e(\mu_{\mathrm{eff}}^2)}{\pi^2 k_{\mathrm{B}}^2}R. \tag{5.1}$$

Here, A_{hf} is a hyperfine coupling constant, N_{A} is Avogadro's number, μ_{B} is Bohr magneton, μ_{eff} is the effective moment, and R is the Wilson ratio. The ΔK_{spin} estimated from this equation was in good agreement with the experimental results. The details of this discussion and comparison of other heavy fermion superconductors are shown in Ref. [5]. In UTe_2, γ_e is about 120 mJ/mol·K^2 and A_{hf} is 51.8 kOe/μ_{B}. It is difficult to obtain the Wilson ratio in heavy fermion materials, but here we adopt 1, which is the usual metal value. Then the expected ΔK is roughly 1%, which is one order of magnitude larger than the observed value. From these results, spin-triplet superconductivity is strongly expected from NMR.

5.2 Change of SC Character in UTe_2 Induced by Magnetic Field

Short Summary

One of the key features of UTe_2 is the enhancement of superconductivity under a magnetic field (H) exceeding 16 T, when H aligns precisely along the b axis. The superconducting (SC) state within this high-field region, termed high-H SC (HHSC), has not been extensively explored, leaving its SC character and spin condition largely ambiguous. Through AC magnetic susceptibility and NMR experiments, this study unveils that the HHSC phase, observed up to 24.8 T, not only bulk feature of UTe_2 but also exhibits considerable sensitivity to the magnetic field's angular orientation. Furthermore, its superconducting nature markedly contrasts with that observed in the low-H SC (LHSC) phase. In the LHSC phase, the prevailing d-vector of the spin-triplet pairing predominantly possesses components along the b axis, yet this orientation shifts, becoming perpendicular to the b axis within the HHSC phase. These findings suggest the emergence of multiple SC phases under the influence of H, attributable to the spin degrees of freedom in the Coopar pair.

5.2.1 *Possible SC State Under Magnetic Field*

UTe_2 crystallizes *Immm* space group (#71, $D_{2\mathrm{h}}$) as discussed in previous chapter. From the group theory, the possible SC symmetry and irreducible representation of spin-triplet superconductivity in a $D_{2\mathrm{h}}$ point group corresponding to the orthorhombic crystal structure of UTe_2 in the zero fields and $H \parallel b$ are listed in Tables 5.1 and 5.2 [6, 7]. A single phase transition can be any of these. For example, if the zero-field

Table 5.1 Classification of the odd-parity SC order parameters for point groups with D_{2h} in a zero field. The irreducible representation (IR) and its basis functions are listed. The dominant spin component in the SC state is also shown

D_{2h} (zero field)		
IR	Basis functions	SC spin comp.
A_u	$k_a\hat{a}$, $k_b\hat{b}$, $k_c\hat{c}$	
B_{1u}	$k_b\hat{a}$, $k_a\hat{b}$	c
B_{2u}	$k_a\hat{c}$, $k_c\hat{a}$	b
B_{3u}	$k_c\hat{b}$, $k_b\hat{c}$	a

Table 5.2 Classification of odd-parity SC phases occurring in UTe_2 under a b-axis magnetic field. The typical order parameters belonging to each IR are listed in Table 5.1

IR of C_{2h} (under field)		
H direction	$A_u^{H\parallel b}$	$B_u^{H\parallel b}$
$H \parallel b$	$A_u + iB_{2u}$	$B_{3u} + iB_{1u}$

transition was A_u, for under the b-axis field, the B_{2u} ($H = 0$) component is induced by the magnetic field and the state is $A_u^{H\parallel b}$. Since this change is a crossover, no phase transition is observed.

5.2.2 AC-Susceptibility Under Magnetic Field

Before performing NMR measurements, AC magnetic susceptibility measurements were performed using an NMR coil to confirm superconductivity. Figure 5.4a shows the variation in $-\Delta\nu/\nu_{\text{tune}}$, as measured by sweeping H at 1.5, 1.0, and 0.6 K. Here, ν is the resonating frequency in the NMR tank circuit. At 1.5 and 1.0 K, the SC transitions were indicated by sudden decreases of $-\Delta\nu/\nu_{\text{tune}}$ in the fields, as shown by the arrows. Although UTe_2 is in the SC state at 0.6 K, $-\Delta\nu/\nu_{\text{tune}}$ was found to exhibit a characteristic H dependence. Elevating H beyond 14 T leads to pronounced reductions in the magnitude of $-\Delta\nu/\nu_{\text{tune}}$; conversely, an increase beyond 16.5 Tesla is associated with an augmenting magnitude of this ratio, signifying the presence of a kink around $H_{\text{kink}} \approx 16.5$ T. Figure 5.4b delineates the temperature (T) dependence of $-\Delta\nu/\nu_{\text{tune}}$ for $\mu_0 H$ values of 7.5, 15.5, 16.5, and 24 Tesla during the cooling process.

The minimum value of $|-\Delta\nu/\nu_{\text{tune}}|$, in relation to SC diamagnetism, was observed at 16 T, consequently demonstrating the same tendency as T_c. This indicates that the HHSC state is bulk properties of UTe_2. The H and T dependencies of $-\Delta\nu/\nu_{\text{tune}}$ suggest that the SC character changed at $\mu_0 H_{\text{kink}} \sim 16.5$ T. In

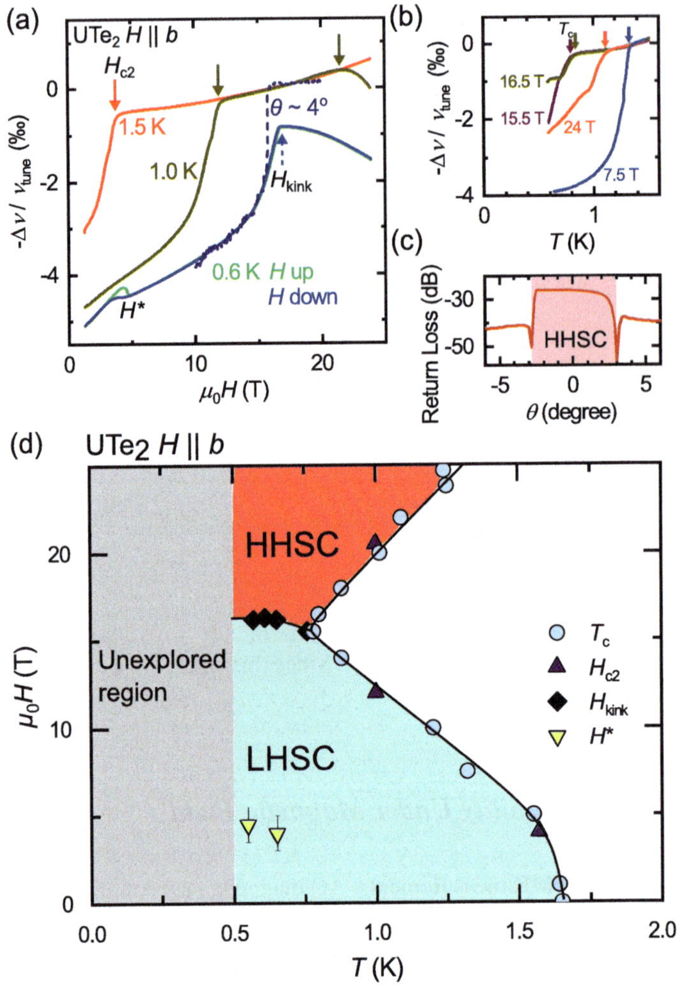

Fig. 5.4 **a** H dependence of $-\Delta\nu/\nu_{\text{tune}}$ values up to 24.8 T, as measured at 0.6, 1.0 and 1.5 K. At 0.6 K, $-\Delta\nu/\nu_{\text{tune}}$ exhibited a kink at H_{kink}; the H dependence of $-\Delta\nu/\nu_{\text{tune}}$, which was determined by performing H-up and H-down sweeps, is shown. The dotted curve shows the H dependence of $-\Delta\nu/\nu_{\text{tune}}$ when the minute θ rotation unexpectedly occurred in the sample. **b** Temperature dependence of $-\Delta\nu/\nu_{\text{tune}}$ in relation to the AC magnetic susceptibility χ_{AC}, as measured at 7.5, 15.5, 16.5, and 24 T. T_c in the field is denoted by the arrow with the same color. **c** Angle dependence of the return loss of the NMR tank circuit at 24 T. When the sample was in the SC state, the quality factor of the circuit Q was lower owing to the change in the impedance of the circuit. High-H superconductivity was observable within $\pm 3°$. **d** SC upper critical field H_{c2} determined by performing T- and H-scan measurements of $-\Delta\nu/\nu_{\text{tune}}$. These figures are reproduced with permission from Ref. [12] © (2023) The American Physical Society

fact, the HHSC state was found to be very sensitive to the angle θ. Additionally, superconductivity was observed within $\pm 3°$ at 24 T, as shown in Fig. 5.4c, and the unexpected minute θ rotation ($\Delta\theta \sim 4°$) that occurred during the experiments completely suppressed the high-H superconductivity, although the SC diamagnetism in the low-H SC (LHSC) state was nearly unchanged, as shown by the dotted curve in Fig. 5.4a. This result is in good agreement with the results presented in previous reports [8, 9]. Based on the T- and H-scan measurements of $-\Delta\nu/\nu_{\rm tune}$, we developed the SC H_{c2} phase diagram shown in Fig. 5.4d. The $H_{\rm kink}$ anomaly, like the overall phase diagram, is consistent with the phase transitions determined by the recent specific-heat measurements [10]. Because the responses to H and θ are different between the LHSC and HHSC states, it is reasonable to consider that the kink at $H_{\rm kink}$ marks a phase transition between the two SC states. Such a transition between two bulk SC states was also confirmed in the work [10] by linear magnetostriction and thermal dilatation, evidencing anomalies due to vortex pinning in both phases. The results of the H-sweep measurement at 0.6 K revealed clear hysteresis behavior at $\mu_0 H^* \sim 4$ T; it was found to be related to an anomaly of the vortex state because the anomaly was not previously observed in the H dependence results for the electronic term in specific-heat measurements [11]. The details of this anomaly have been needed to be studied.

5.2.3 *NMR Study of LHSC and HHSC State*

To examine the superconducting (SC) characteristics, particularly spin susceptibility in the high-HHSC state, we conducted ^{125}Te nuclear magnetic resonance (NMR) studies at the Te(2) peak, with a higher Knight shift (K). Figure 5.6a and b show the Te(2) NMR spectra measured at various temperatures below 2.5 K at 1 and 24 T, respectively. At 1 T, the single-peak spectrum gradually shifted to the low-K side in the normal state and sharply shifted immediately below T_c; this was accompanied by spectrum broadening. The ^{125}Te NMR spectrum measured under conditions of 24 T and 2.5 K revealed a double-peak structure that is attributable to its high resolution; the right peak was found to have a 0.04% larger K than the main peak. The H dependence of the Te(2)-NMR spectrum is shown in Fig. 5.5. Several possibilities were considered for the origin of the larger-K peak; they include the occurrence of a mosaic structure and/or minute U-atom deficiency in the single-crystal sample. In the former case, the misalignment of the mosaic was estimated to be 2.0° (8.1°) on the a (c) axis; additionally, the ^{125}Te NMR measurement for the higher-T_c single crystal is critical for the latter possibility because T_c seems to be very sensitive to a U-atom deficiency [7, 13, 14]. Further experiments are required to clarify the origin of the larger K peak. As T was decreased, the two peaks gradually shifted to the lower-K side in the same manner. Because the resolution of the higher-K peak is not sufficient for analysis, we focus on the main peak shown by the arrows.

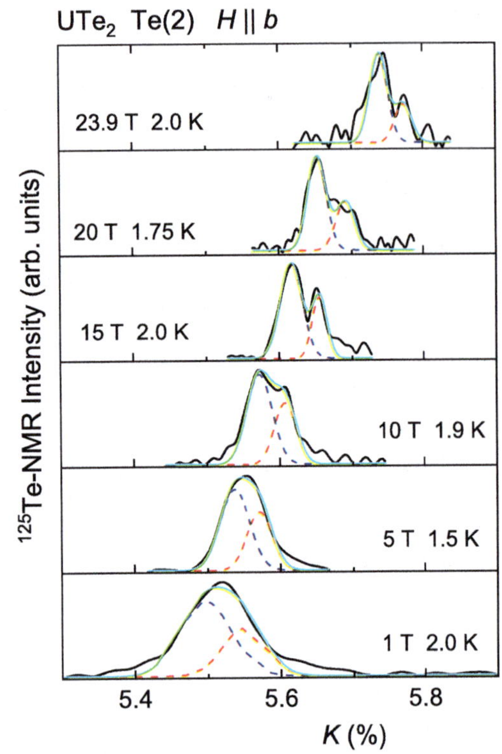

Fig. 5.5 Te(2)-NMR spectrum measured at 1, 5, 10, 15, 20, and 24 T. It is shown that the two-peak structure becomes clear with increasing the fields. Since the peak position of the sub-peak was not precisely determined, we adopted the peak position of the main peak and analyzed it. These figures are reproduced with permission from Ref. [12] © (2023) The American Physical Society

Figure 5.6c shows the temperature dependence of K of the main peak, as determined from the NMR spectra measured at 1, 10, 15, 20, and 24 T. A decrease in K was clearly observed at T_c for 1 and 10 T; the magnitude and H dependence of the K decrease below T_c are in agreement with previous results [2, 15]. In contrast, when 20 and 24 T were applied, K gradually decreased at temperatures below 2.5 K without any appreciable anomaly at $T_c(H)$. To quantify the Knight-shift decrease (ΔK) ascribed to the superconductivity, the normal-state T dependence was subtracted from the observed K_b and ΔK was plotted for each H, as shown in the inset of Fig. 5.6c. ΔK was near zero in the SC state for values above 15 T, although this field is still in the LHSC. This behavior is consistent with previous NMR measurements [2, 15]. A similar $\Delta K \sim 0$ trend was previously observed in the a-axis Knight-shift measurement results for the LHSC state, where the dominant SC spin component occurred along the a axis [3]. Thus, these results indicate that b-axis spin-polarized superconductivity is induced by a b-axis magnetic field.

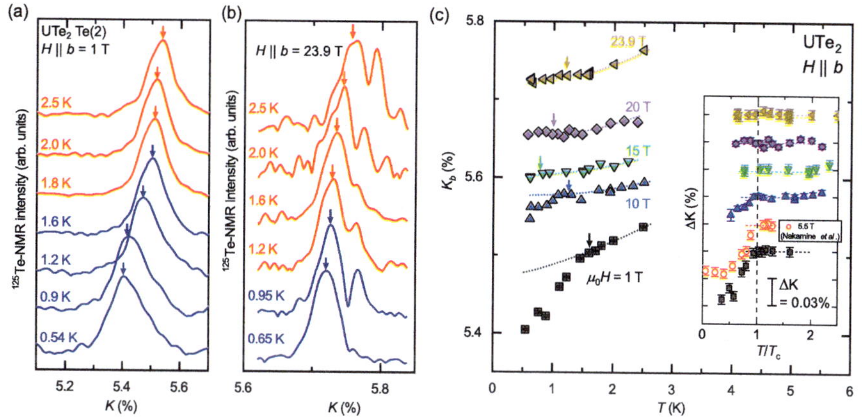

Fig. 5.6 Te(2) NMR peaks measured at various temperatures below 2.5 K at ~1 T (**a**) and ~24 T (**b**). **c** Temperature dependence of K, as determined by analyzing the Te(2) NMR peak shown by the arrow. The dotted line is the normal-state behavior extrapolated using the 2nd-order polynomial function as explained in the supplemental materials. (inset) Temperature dependence of the change in the Knight shift from the normal state. The temperature dependence of the normal-state Knight shift was fitted and subtracted using a quadratic function. The horizontal dashed line represents the baseline, and the vertical dashed line represents the transition temperature. These figures are reproduced with permission from Ref. [12] © (2023) The American Physical Society

5.2.4 Possible SC Phase in HHSC State

We will now discuss possible SC states in the HHSC region. In light of the observed spin-susceptibility and field-boosted behavior, the ground state of the HHSC is $A_{\mathrm{u}}^{H\|b}$, as outlined in Table 5.2. This is because the SC spin component is parallel to $H \parallel b$ in the HHSC region. Though there is a smooth transition in ΔK, the distinct kink anomaly observed in the field-dependent $-\Delta\nu/\nu_{\mathrm{tune}}$ signifies a phase transition between the high-HHSC and low-HHSC states. Consequently, the low-HHSC state is classified as $B_{\mathrm{u}}^{H\|b}$. These results strongly support the $B_{3\mathrm{u}}$ scenario at a low-field limit [1–3, 15]. Under $H \parallel b$, $B_{3\mathrm{u}}$ at zero field becomes $B_{\mathrm{u}}^{H\|b}$ with crossover (without any transition). It is important to note that thermodynamic principles restrict that it is not allowed the "tri-critical point" with three 2nd order phase transition lines. Thus, the phase transition line inside the SC region should be 1st order or there is a hidden phase transition line with 2nd order phase transition [6]. This hidden phase transition line has yet to be confirmed, but has been suggested as a possibility by Sakai *et al* from magneto-resistance measurements [16]. The results of the crude up-down H-sweep measurement of $-\Delta\nu/\nu_{\mathrm{tune}}$ at 0.6 K revealed the occurrence of one kink without any hysteresis near H_{kink} [Fig. 5.4b]; this seems to exclude the "1st order" phase transition scenario. Rather we suggest the presence of another phase transition line characterized by the SC properties of the HHSC state such as $\Delta K = 0$. Further precise NMR measurements are required to understand the relationship between the LHSC and HHSC phases. The suggested phase diagram is shown in Fig. 5.7.

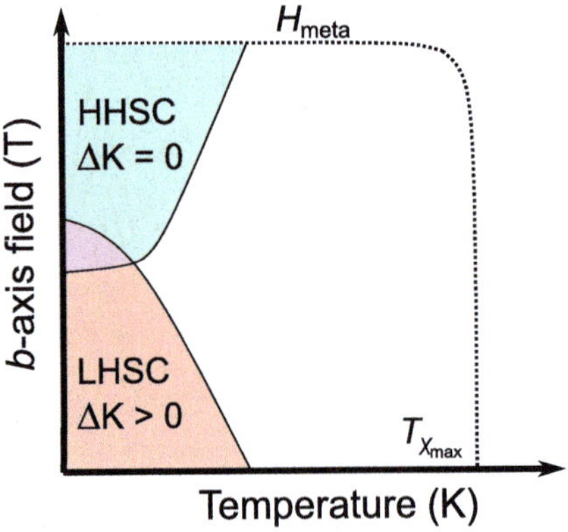

Fig. 5.7 Possible phase diagram suggested from our NMR results

In addition, it is noteworthy that the enhancement of the superconductivity against H was found to be stronger than that previously reported [8, 17, 18]. Because the value of T_c at $H = 0$ for the current sample (1.67 K) was slightly higher than that of previous samples ($\sim$1.5 K), the upturn behavior is seemingly dependent on the sample quality, suggesting the intrinsic properties of UTe_2. A similar level of superconductivity robustness by $H \parallel b$ was observed in the FM superconductors URhGe [19] and UCoGe [20], in which critical FM fluctuations were determined to play an important role [21–24]. Because superconductivity occurs in the paramagnetic state of UTe_2, such critical FM fluctuations were not anticipated. Alternatively, we speculate that the critical fluctuations related to the incommensurate antiferromagnetic fluctuations [25, 26], which may be induced by $H \parallel b$ above 16.5 T, plays an important role in the mechanism governing the HHSC state. It is interesting that the SC pairing interaction can be tuned by adjusting the H applied along the b axis (magnetic hard axis); this seems to be a common feature of U-based FM and nearly FM superconductors with Ising anisotropy under normal-state magnetic conditions, although the SC pairing interaction is not clarified in UTe_2.

5.3 Drastic Change in Magnetic Anisotropy of UTe_2 Under Pressure Revealed by ^{125}Te-NMR

Short Summary

To investigate the normal-state magnetic properties of UTe_2 under pressure, we perform ^{125}Te nuclear magnetic resonance (NMR) measurements up to 2 GPa. Below 1.2 GPa, the Knight shift along the b-axis exhibits a broad peak at a temperature designated as $T_{\chi max}$ on cooling, consistent with findings from magnetization measurements under pressure. $T_{\chi_{max}}$ decreases with increasing pressure and disappears at the critical pressure $P_c = 1.7$ GPa, above which superconductivity is destroyed.

This trend is similarly observed in the temperature dependency of the nuclear spin-lattice relaxation rate, $1/T_1$. At lower pressures, $1/T_1$ exhibits standard Fermi-liquid behavior ($1/T_1T$ = constant) at low temperatures, indicating the formation of the heavy-fermion state. Above P_c, $1/T_1T$ follows a $1/T$ behavior without any crossover to the heavy-fermion state down to the lowest temperature (~3 K). In addition, the NMR signals disappear below 3 K, due to the influence of the magnetically ordered moments. From the pressure dependence of $T_{\chi_{max}}$ and Knight shift, it was found that the Fermi surface character is abruptly changed at P_c, and that superconductivity is observed only in the heavy-fermion state.

5.3.1 *Superconducting Transition Temperature Under Pressure*

First, we performed AC susceptibility measurements under pressure to reproduce the superconducting transition under pressure. Figure 5.8a shows the T dependence of the AC magnetic susceptibility (χ_{AC}) under various pressures, and T_c's are shown by arrows. T_c first increases and reaches a maximum at approximately 1.2 GPa with increasing pressure, and then superconductivity suddenly disappears above P_c. The P dependence of the present T_c is in good agreement with previous results [27, 28], and the P_c of this sample was determined to be ~1.7 GPa. This is because the sharp transition ascribed to the bulk Meissner signal was not observed, although a relatively small decrease in χ_{AC} was observed at 1.7 GPa. We considered that this result is due to the first-order transition-like behavior described below, which

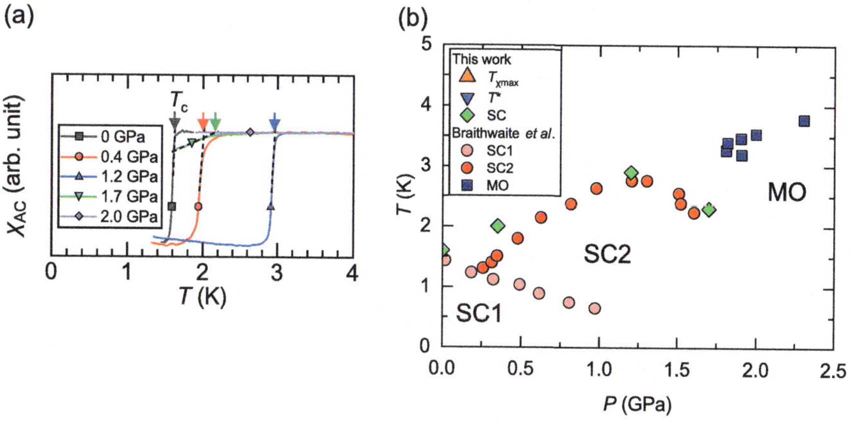

Fig. 5.8 **a** T dependence of magnetic AC susceptibility χ_{AC} at several pressures. The measurement was performed using an NMR tank circuit on cooling. T_c's are shown by the arrows. **b** P–T phase diagram of UTe_2

Table 5.3 The hyperfine coupling constant ($A_{\rm hf}$) of UTe_2 at several pressures. $A_{\rm hf}^{1[2]}$ is the hyperfine coupling constant of Te(1) [Te(2)]. $A_{\rm hf}$ of Te(1) and Te(2) for $H \parallel c$ has almost same value and are indistinguishable

$H \parallel b$		
P (GPa)	$A_{\rm hf}^1$ (T/μ_B)	$A_{\rm hf}^2$ (T/μ_B)
0	4.1 ± 0.3	5.6 ± 0.4
1.2	4.3 ± 0.3	5.6 ± 0.4
1.7	4.4 ± 0.5	5.7 ± 0.6
$H \parallel c$		
P (GPa)	$A_{\rm hf}$ (T/μ_B)	
0	3.9 ± 0.2	
1.2	4.0 ± 0.2	
1.7	4.4 ± 0.3	

results in a small superconducting volume fraction. No Meissner signal was observed at $P = 2.0$ GPa. As shown in Fig. 5.8b, our results almost reproduce the previous specific heat measurements [27].

5.3.2 Magnetic Anisotropy of Normal State Susceptibility Under Pressure

We found a drastic change in the T dependence of K_b around P_c. First, to see how the magnetic field created by the electrons at the nuclear position changes, a relationship between K and χ, called a K-χ plot, is shown in the inset of Fig. 5.9. Usually, the NMR Knight shift and the susceptibility χ follow the relation, $K = A_{\rm hf}\chi + b\chi_0$, where $A_{\rm hf}$ is the hyperfine coupling constant and $b\chi_0$ term is the temperature-independent background contribution. In this case, the background contribution mainly comes from the bulk susceptibility of the pressure cell. In fact, good linear relations between the two quantities were confirmed down to the lowest temperature beyond a broad maximum in the previous ambient-pressure susceptibility [31]. The hyperfine coupling constant estimated from the slope of the K-χ plot is summarized in Table 5.3, and is almost P-independent; the P dependence of the NMR physical quantities is reasonably ascribed to that of the magnetic-susceptibility properties. Now that it is clear that K reflects the magnetic susceptibility, we focus on the temperature dependence of K. With increasing pressure, the b-axis $T_{\chi_{\rm max}}$ gradually shifts to a low temperature, which is consistent with the results of the AC and DC magnetic susceptibility measurements [32]. Above P_c, K_b does not show the broad maximum and diverges to the lowest temperature.

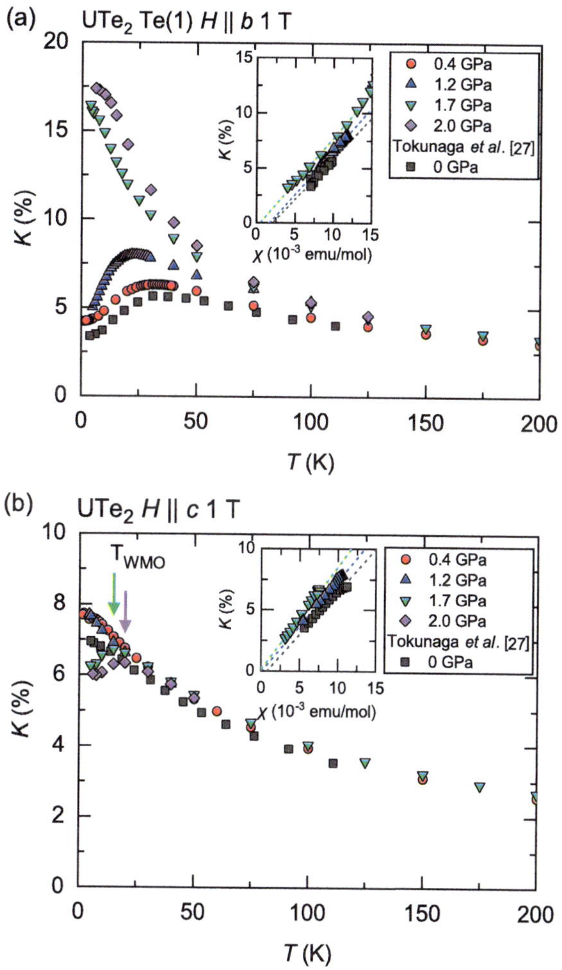

Fig. 5.9 Temperature dependence of the NMR Knight shift K for **a** $H \parallel b$ and **b** $H \parallel c$. Below 1.7 GPa, i.e., P_c, K_b shows a broad maximum in T dependence. In contrast, above P_c, K increases with the decreasing temperature down to the lowest temperature. (inset) K-χ plots for 0, 1.2, and 1.7 GPa for **a** $H \parallel b$ and **b** $H \parallel c$. The value of χ under pressure was reported in Ref. [29]. The dashed lines in the figure are the linear-fitting line. These figures are reproduced with permission from Ref. [30] © (2023) The American Physical Society

In contrast, the absolute value of K_c is almost the same at all pressures, and thus the magnetic hardest axis at low temperatures changes from the b to c axis above P_c. In addition, an appreciable peak in K_c appears in the temperature dependence above P_c. It was reported that, above P_c (1.7 GPa), the magnetic susceptibility along the c-axis (χ_c) shows a broad maximum at $T_{WMO} \sim 10$ K, where a magnetic short-range order develops [29]. Thus, the peak in K_c corresponds to the peak in χ_c at T_{WMO}, and this gradually shifts to a higher temperature with increasing pressure. These results indicate that the origin of the peak in K_c is different from that in K_b at $T_{\chi_{max}}$.

5.3.3 Pressure Dependence of Spin Fluctuations

Next, focus on the relaxation measurements to see the magnetic fluctuations. Figure 5.10a and b show the T dependence of $1/T_1T$, which is proportional to the q-averaged dynamical spin susceptibility in (a) $H \parallel b$ and (b) $H \parallel c$. In contrast to

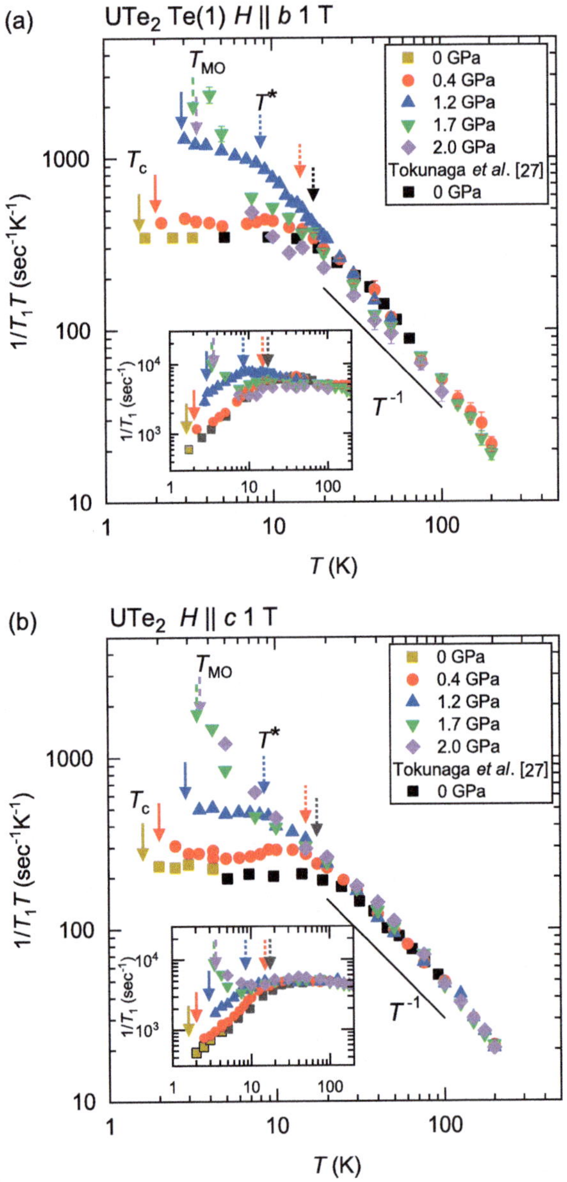

Fig. 5.10 T dependence of the nuclear spin–lattice relaxation rate divided by T ($1/T_1T$), which is proportional to the q-averaged dynamical spin susceptibility in **a** $H \parallel b$ and **b** $H \parallel c$. The insets are the T dependence of $1/T_1$ for $H \parallel b$ and $H \parallel c$. T_c, T_{MO}, and T^* are shown by solid, dashed, and dotted arrows, respectively. These figures are taken from Ref. [30]

the Knight shift, the T and P dependences of $1/T_1T$ are almost the same in both directions, although $1/T_1T$ behavior below T^* is slightly different at 1.2 GPa. This is because $1/T_1T$ probes magnetic fluctuations perpendicular to the applied magnetic field and both $1/T_1T$'s are governed by fluctuations arising from the magnetic easy-axis (a-axis) component [31]. At ambient pressure, $1/T_1T$ exhibits the typical heavy-fermion behavior as observed in UPt_3 [33]. $1/T_1T$ is proportional to $1/T$ in the high-temperature region owing to the localized $5f$-electron character. On cooling, $1/T_1T$ deviates from the $1/T$ behavior owing to the hybridization between the localized $5f$-electron and conducting U-6d and Te-5p electrons [7]. The U-5f electrons show an itinerant character below the Kondo coherent temperature characterized by $T_{\chi_{\max}}$, and $1/T_1T$ becomes almost constant below T^*, which is a characteristic behavior of the heavy-fermion state. When pressure is applied, the $T_{\chi_{\max}}$ decreases, and the Korringa value (the constant value of $1/T_1T$ at low T) increases, indicating that the bandwidth becomes narrower and the electron correlation becomes stronger. In contrast, $1/T_1T$ continues to increase down to the lowest T above P_c, analogous to the K_b, and the heavy-fermion state was no more observed down to 3 K. It indicates that the heavy-fermion state and the Kondo coherent state were destroyed above P_c.

From the present results, as well as those previously reported [27–29, 32], we developed the P–T phase diagram, as shown in Fig. 5.4b. This phase diagram shows that the Kondo coherent state was destroyed above P_c, where the U 5-f state remained localized down to 3 K. It was clarified that superconductivity is observed only in the Kondo coherent state, where the U 5-f state becomes itinerant. We point out the similarity between the P dependence and $H \parallel b$ phase diagrams because the superconductivity is enhanced by the application of P or $H \parallel b$ but is abruptly suppressed by the destruction of the Kondo coherent state in both cases. Above P_c or metamagnetic field, magnetic properties are not well-understood so far. The gradual increase in K_b at low T by applying a small pressure [Fig. 5.4a] is the same tendency as the H dependence of K_b, in which K_b just above T_c slightly increases with increasing H up to 15 T [15]. It is considered that the gradual increase in K_b and gradual decrease in $T_{\chi_{\max}}$ with pressure lower the metamagnetic fields in $H \parallel b$, as was actually observed in the high-field magnetization measurement under pressure [18]. In addition, K_b at low T discontinuously increases at P_c, suggesting that the evolution of the electronic state at P_c looks like a first-order transition.

5.3.4 *P–T* Phase Diagram

From the experimental results, we have made the P-T phase diagram as shown in Fig. 5.11. Above P_c, the NMR signals of Te(1) and Te(2) disappeared below T_{MO} assigned by the resistivity and magnetization measurements [29, 37] because of the influence of the magnetic ordered moments. Recently, inelastic neutron scattering measurements reported that the magnetic fluctuations along the a axis with $Q_{INS} = (0, 0.57, 0)$ are roughly proportional to $1/T_1T$ at ambient pressure [25, 26]. If this relation is still preserved even above P_c, $1/T_1T$ diverges toward T_{MO}, indicating that

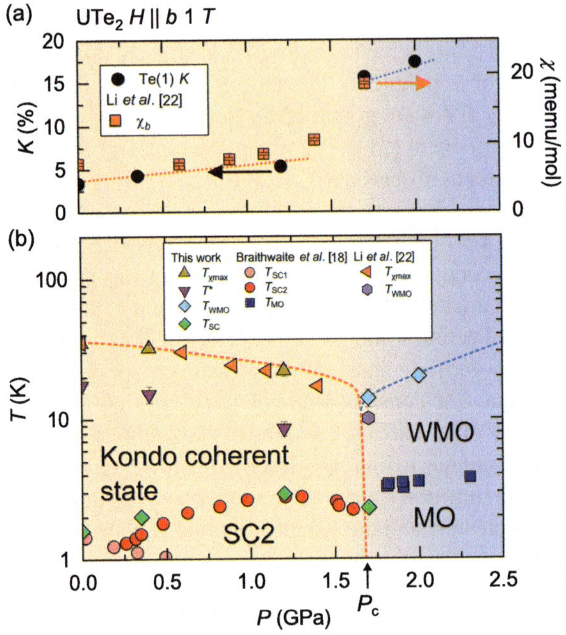

Fig. 5.11 **a** P dependence of the NMR Knight shift at 6 K. The value of the Knight shift suddenly increases around 1.7 GPa with increasing pressure. **b** P–T phase diagram of UTe_2. T_{SC} is the superconducting critical temperature. $T_{\chi_{max}}$ is the temperature peak position in the b-axis Knight shift and T^* is the crossover temperature, below which $1/T_1T$ is almost constant. $T_{\chi_{max}}$ and T^* decrease with increasing pressure and suddenly disappear above P_c. T_{WMO} is the peak temperature in the c-axis Knight shift and χ. These figures are reproduced with permission from Ref. [30] © (2023) The American Physical Society

the MO state is the AFM ordered state with incommensurate Q_{INS}. In such a case, the NMR spectrum is smeared out due to distributed internal fields at the Te sites.

5.4 Superconducting-Spin Rotation in Multiple Superconducting Phases of UTe_2

Short Summary

UTe_2 is now known to be a spin-triplet topological superconductor and has multiple SC phases under magnetic fields and pressure. However, SC character under these conditions is unknown. In this section, our nuclear magnetic resonance and AC susceptibility measurements on single crystalline UTe_2 under pressure revealed the SC spin state in the multiple SC phases of UTe_2. We found that SC spin states at ambient pressure and under pressure are different and that the further change of the spin state occurs at a low temperature under pressure. The SC spin rotation, associated with SC-SC phase transition, is indisputable evidence of the spin-triplet SC phenomena. Our work on UTe_2 paves the way for studying the spin-triplet multiple SC phases.

5.4.1 NMR Spectrum Under Pressure

In this measurement, we focused on 1.2 GPa, i.e., the pressure at which the transition temperature reaches a maximum of 3 K, to clarify the behavior of the SC2 phase. The T variation of the NMR spectrum of Te(2) at $P = 1.2$ GPa is shown in Fig. 5.12a, and the T dependence of the full width at half maximum (FWHM), the NMR Knight shift (K) determined with the NMR spectrum, and χ_{AC} is shown in Fig. 5.12b. As shown in previous section and papers [29, 30], K_b exhibits a broad maximum at $T_{\chi max}$ and gradually decreases with decreasing T, similar to the T dependence of χ. Owing to the increase in T_c and the decrease in $T_{\chi max}$ in $P > 0.3$ GPa [27, 29], K_b highly depends on T around T_c at 1.2 GPa unlike the case of ambient pressure (0 GPa). Therefore, although the absolute value of the spin part in K (K_{spin}) decreases with decreasing T, $\Delta K \equiv K_{spin}^{SC} - K_{spin}^{normal}$ is almost zero above 0.5 K as shown in Fig. 5.13a, indicating that the spin susceptibility in SC2 is the same as that in the

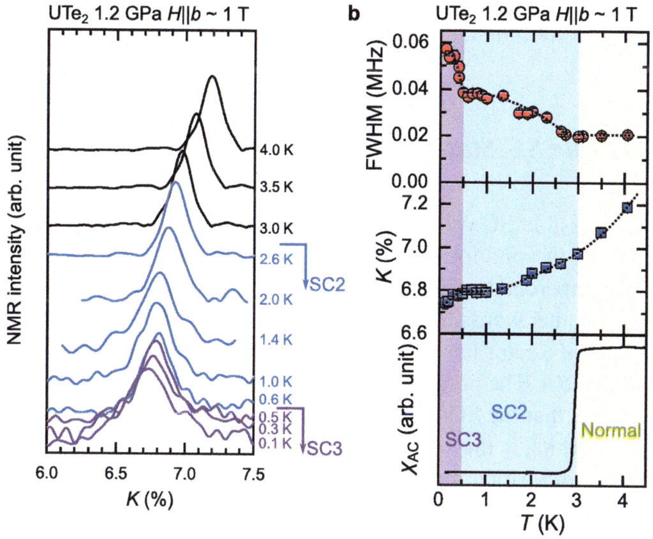

Fig. 5.12 **a** Temperature variation of the NMR spectra of UTe_2 at 1.2 GPa. **b** T dependence of the FWHM, K of the Te NMR spectrum (middle), and AC χ at 1.2 GPa. These figures are reproduced with CC-BY 4.0 license from Ref. [34] (2023)

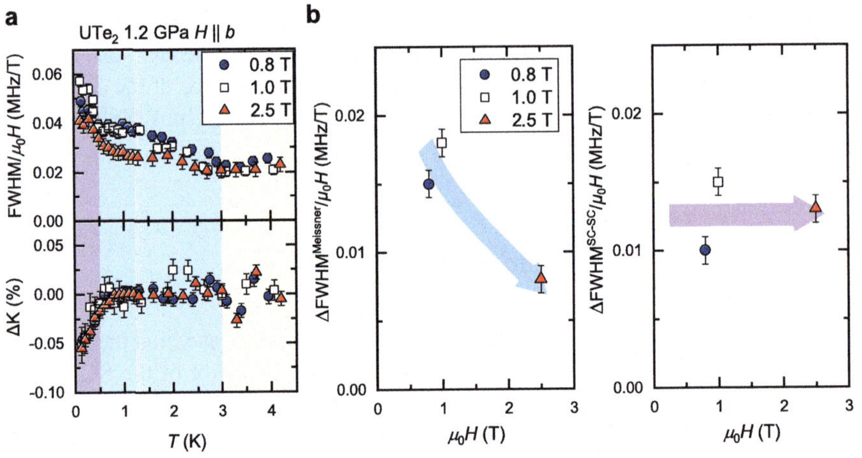

Fig. 5.13 **a** Temperature dependence of (Top) the FWHM/$\mu_0 H$ and (Bottom) ΔK at several fields of the Te NMR spectrum. **b** The field dependence of the ΔFWHM/$\mu_0 H$ in the SC2 and SC3 states. These figures are reproduced with CC-BY 4.0 license from Ref. [34] (2023)

normal state for $H \parallel b$. Here, $K_{\rm spin}^{\rm normal}$ ($K_{\rm spin}^{\rm SC}$) is $K_{\rm spin}$ in the normal (SC) state, and is estimated from the extrapolation above $T_{\rm c}$. Below 0.5 K, we observed a sudden decrease in K_b and a broadening of the NMR spectrum, although no additional change was observed in $\chi_{\rm AC}$. The magnitude of the decrease in K_b below 0.5 K is almost the same as that in K at 0 GPa [1]. This is further evidence of the occurrence of the phase transition inside the SC state from the microscopic point of view.

5.4.2 *Possible SC State Under Pressure*

We consider possible SC order parameters suggested by the present NMR results. As discussed in the previous papers [2, 3] and this chapter, the order parameter of spin-triplet superconductivity is $\boldsymbol{d}$-vector, which is perpendicular to the SC spin components. In other words, K_i proportional to the spin susceptibility along the i axis decreases when $\boldsymbol{d}$-vector has the i component, but is unchanged when the $\boldsymbol{d}$-vector has no i component. The possible SC phases at zero field are listed in Table 6.1. Our findings indicate that the SC-spin is oriented to the b axis in SC2, and rotates to the a axis in SC3, which is the same direction as that of SC1. This is SC-spin rotation, evidencing the spin-triplet multiple SC phases with spin degrees of freedom.

An additional key observation from Fig. 5.12b is that the FWHM of the Te(2) site shows an additional increase below 0.5 K. Figure 5.13a shows the temperature dependence of the FWHM divided by the applied fields ($\mu_0 H$), which reflects the distribution of the spin-susceptibility, and ΔK measured at the Te(2) site in 0.8, 1.0, and 2.5 T. In general, the FWHM of the NMR spectrum increases below $T_{\rm c}$

due to SC diamagnetism. This effect was observed below 3.0 K in SC2 at both the Te sites and was suppressed with increasing $\mu_0 H$, as shown in Fig. 5.13b. This behavior is expressed by the SC diamagnetic effect. However, the additional increase in the FWHM below 0.5 K cannot be explained by the SC diamagnetic effect, since χ_{AC} shows no anomaly around 0.5 K, and the additional increase below 0.5 K is proportional to the applied field, as shown in Fig. 5.13c. This field dependence was not observed in the broadening of the FWHM in SC1, although the SC diamagnetic effect was observed. In addition, the ratio between the FWHM of the Te(1) and Te(2) spectra in SC3 is almost unchanged from that in the normal state. These results indicate that the additional linewidth broadening in SC3 arises from the inhomogeneity of the spin susceptibility, which is a different character from SC1.

5.5 Line Broadeing in SC3 State

Several possibilities are pointed out for the origin. One possibility is due to the SC vortex state in SC3. Since spin susceptibility is unchanged in SC2, the NMR signal from the vortex core is observed at almost the same position as the SC NMR signal outside the vortex. On the other hand, the spin susceptibility decreases in SC3, and the signal from the vortex core should be observed at a different position from the SC NMR signal, resulting in the NMR signal becoming broader and anisotropic in the mixed state of SC3. However, this possibility would be unlikely, because the measurement field is far below H_{c2} in $H \parallel b$ [32], and thus the vortex region would be negligibly small.

Another possibility is due to the inhomogeneity of the sample. For example, given that the transition temperatures in the sample are somewhat spread out and that the superconducting transition occurs continuously, the line width may become broader. However, it is hard to explain the obtained spectra from this scenario. For the SC1 spectrum, the SC-state spectrum (0.5 K) is clearly shifted toward the low-Knight shift side compared to the normal state spectrum (1.6 K) (Fig. 5.14a). In contrast, in the SC3 spectrum measured at the same magnetic field, the peak of the spectrum is shifted to the low K side, but the signal is also observed at the high K side, and the spectral width is broadened (Fig. 5.14b).

To consider this line width broadening more quantitatively, we have performed the following simulations. As a simple assumption, we consider a two-phase coexistence, which means that a part of the sample remains in the SC2 phase and a part in the SC3 phase. This case can be easily ruled out from the following comparison of the shape of the spectrum changes. Figure 5.15a shows a simulation of what happens to the spectrum in the case of simple phase separation. The experimental spectrum cannot be reproduced by two Lorentz functions.

Next, we consider the inhomogeneity in the sample. Figure 5.15b and c show a simulation for the 0.13 K spectrum assuming the normal distribution of the 0.7 K spectrum. The simulation in (b) assumes a Gaussian function with $\sigma = 0.02$ as the distribution of the center and intensity of the Lorentz function, and in (c) a Gaussian

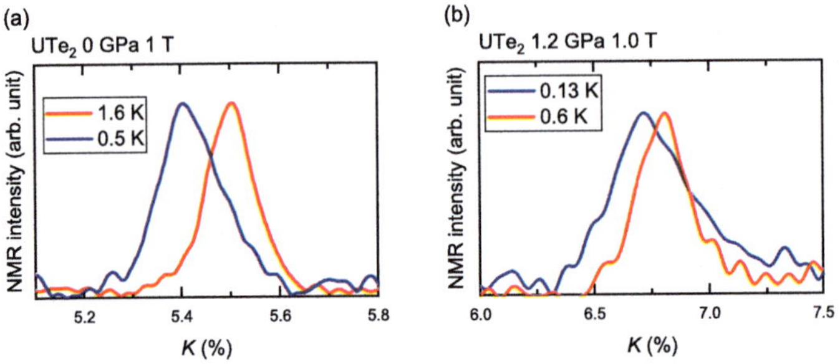

Fig. 5.14 **a** Change in ^{125}Te-NMR spectra in the SC1 phase. **b** Change in ^{125}Te-NMR spectra in the SC3 phase; after the SC3 transition, the spectrum broadened to the direction with a larger Knight shift than in the normal state. These figures are reproduced with CC-BY 4.0 license from Ref. [34] (2023)

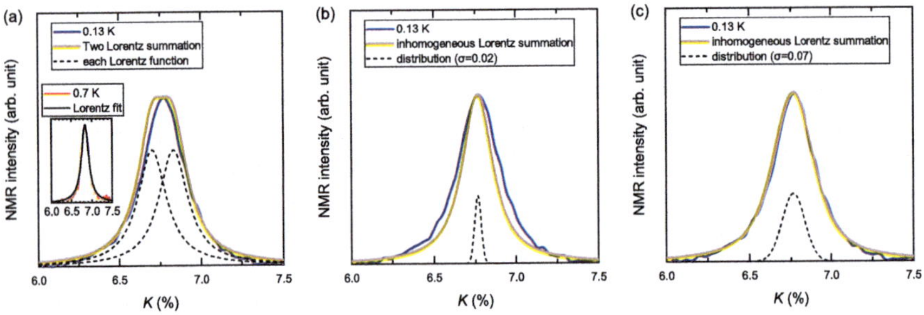

Fig. 5.15 **a** The NMR spectrum of the SC3 phase and the spectrum expected when the SC1 and SC2 phases are phase-separated in real space (Two-Lorentz summation). In the model of phase separation, an equal amount of mixing would result in a flat-top, anisotropic spectrum depending on the fraction of either phase, which is different from a symmetrically broad spectrum as obtained in the experiment. The inset is the spectrum just above the SC3 transition used as a model for the Lorentz function. **b**, **c** Simulations for the 0.13 K spectrum assuming the normal distribution of the 0.7 K spectrum. **b** Spectrum was calculated with normal distribution with $\sigma = 0.02$. **c** Spectrum was calculated with normal distribution with $\sigma = 0.07$. These figures are reproduced with CC-BY 4.0 license from Ref. [34] © (2023)

function with $\sigma = 0.07$. The Gaussian function with $\sigma = 0.02$ has a small intensity of about 6% of the center at the position of the SC2 phase signal, and the distribution of the Knight shift higher than SC2 is negligible. However, the spectrum reproduced by this distribution is sharper than observed. Conversely, the shape of the spectrum is matched in Fig. 5.15c. In this case, the spectrum is almost reproduced, but its distribution is physically strange because the weight of the Knight shift, which is higher than that of SC2, is quite large. This means that a simple Gaussian distribution cannot explain the broadening of the linewidth.

These results indicate that, whatever the inhomogeneities in the sample, there is an intrinsic linewidth broadening, associated with the SC-SC transition, which is proportional to the magnetic field. In other words, it represents an increase in the distribution of the magnetic susceptibility with the SC-SC transition. At present we have no clear interpretation of this phenomenon, but we believe that we may have observed an effect due to the mixing of the parameters of the two superconducting order parameters.

5.5.1 Similarity Between *P* and H_b Effect

As discussed in the previous chapter for the normal state under pressure, the normal-state properties suddenly change just above P_c, and the P–T phase diagram and the H–T phase diagram are similar in terms of the Kondo coherent state [30]. We found further similarity between the SC2 and HFSC states since K_b is unchanged in both SC states.

Figure 5.16 shows the schematic image of the T vs H along the b axis (H_b) phase diagram of UTe_2 at several pressures by taking account of the previous experimental results [2, 8, 34]. At ambient pressure, $T_c(H)$ decreases in the LFSC (Low field

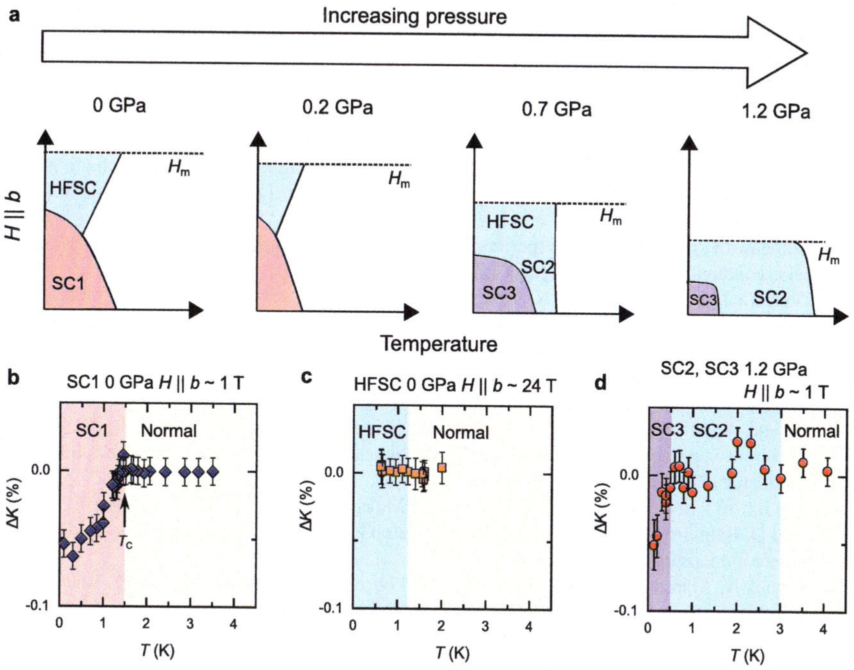

Fig. 5.16 **a** The schematic images of the H–T phase diagram of UTe_2 under several pressures. The bold lines are the transition lines determined by the specific heat measurements. The dashed line represents the metamagnetic transition line. **b–d** The temperature dependence of ΔK in the 4 states is shown in Fig. 5.16a. Part of these figures are reproduced with CC-BY 4.0 license from Ref. [34] © (2023)

superconducting) state, and increases in the HFSC (High field superconducting) state up to the metamagnetic field 35 T (H_{me}). In the LFSC state, K_b slightly decreases below T_{c} [2] (see Fig. 5.16b). As mentioned above, K_b is unchanged in the HFSC state [34] as shown in Fig. 5.16c, indicating that the $\boldsymbol{d}$-vector has no $\hat{b}$ component. At an intermediate pressure $\sim$0.2 GPa, it is expected that H_{m} decreases and the L-shape field dependence of T_{c} remains. At the pressure $\sim$1.2 GPa where T_{c} becomes maximum, H_{m} is suppressed to almost half of H_{m} at $P = 0$ and there is no more L-shape behavior in $H_b - T$ phase diagram. Considering the similarity between the HFSC and SC2 states as shown in Fig. 5.16d, the HFSC, and SC2 states merge and cover SC3. On the basis of the image, we suggest that the spin susceptibility along the b axis (χ_b) would be a key parameter in both phases since χ_b increases with H_b [15, 34–36] and P up to H_{m} and P_{c} [29, 30].

References

1. Nakamine G, Kitagawa S, Ishida K, Tokunaga Y, Sakai H, Kambe S, Nakamura A, Shimizu Y, Homma Y, Li D, Honda F, Aoki D (2019) Superconducting properties of heavy fermion UTe2 revealed by 125Te-nuclear magnetic resonance. J Phys Soc Jpn 88:113703
2. Nakamine G, Kinjo K, Kitagawa S, Ishida K, Tokunaga Y, Sakai H, Kambe S, Nakamura A, Shimizu Y, Homma Y, Li D, Honda F, Aoki D (2021) Anisotropic response of spin susceptibility in the superconducting state of UTe2 probed with 125Te-NMR measurement. Phys Rev B 103:L100503
3. Fujibayashi H, Nakamine G, Kinjo K, Kitagawa S, Ishida K, Tokunaga Y, Sakai H, Kambe S, Nakamura A, Shimizu Y, Homma Y, Li D, Honda F, Aoki D (2022) Superconducting order parameter in UTe2 determined by knight shift measurement. J Phys Soc Jpn 91:043705
4. in preparation
5. Kotegawa H (2019) Toward experimental determination of spin-triplet pairing in new exotic superconductor UTe2. JPSJ News Comm 16:16
6. Ishizuka J, Sumita S, Daido A, Yanase Y (2019) Insulator-metal transition and topological superconductivity in UTe2 from a first-principles calculation. Phys Rev Lett 123:217001
7. Aoki D, Brison J-P, Flouquet J, Ishida K, Knebel G, Tokunaga Y, Yanase Y (2022) Unconventional superconductivity in UTe2. J Phys: Condens Matter 34:243002
8. Knebel G, Knafo W, Pourret A, Niu Q, Vališka M, Braithwaite D, Lapertot G, Nardone M, Zitouni A, Mishra S, Sheikin I, Seyfarth G, Brison J-P, Aoki D, Flouquet J (2019) Field-Reentrant Superconductivity Close to a Metamagnetic Transition in the Heavy-Fermion Superconductor UTe2. J Phys Soc Jpn 88:063707
9. Ran S, Liu I-L, Eo YS, Campbell DJ, Neves PM, Fuhrman WT, Saha SR, Eckberg C, Kim H, Graf D, Balakirev F, Singleton J, Paglione J, Butch NP (2019) Extreme magnetic field-boosted superconductivity. Nat Phys 15:1250–1254
10. Rosuel A, Marcenat C, Knebel G, Klein T, Pourret A, Marquardt N, Niu Q, Rousseau S, Demuer A, Seyfarth G, Lapertot G, Aoki D, Braithwaite D, Flouquet J, Brison J-P (2022) Thermodynamic evidence for two superconducting phases under magnetic field in UTe2. arXiv:2205.04524
11. Kittaka S, Shimizu Y, Sakakibara T, Nakamura A, Li D, Homma Y, Honda F, Aoki D, Machida K (2020) Orientation of point nodes and nonunitary triplet pairing tuned by the easy-axis magnetization in UTe2. Phys Rev Res 2:32014
12. Kinjo K, Fujibayashi H, Kitagawa S, Ishida K, Tokunaga Y, Sakai H, Kambe S, Nakamura A, Shimizu Y, Homma Y, Li DX, Honda F, Aoki D, Hiraki K, Kimata M, Sasaki T (2023) Change of superconducting character in UTe2 induced by magnetic field. Phys Rev B 107:L060502

13. Rosa PFS, Weiland A, Fender SS, Scott BL, Ronning F, Thompson JD, Bauer ED, Thomas SM (2022) Single thermodynamic transition at 2 K in superconducting UTe2 single crystals. Commun Mater 3:33
14. Haga Y, Opletal P, Tokiwa Y, Yamamoto E, Tokunaga Y, Kambe S, Sakai H (2022) Effect of uranium deficiency on normal and superconducting properties in unconventional superconductor UTe2. J Phys: Condens Matter 34:175601
15. Nakamine G, Kinjo K, Kitagawa S, Ishida K, Tokunaga Y, Sakai H, Kambe S, Nakamura A, Shimizu Y, Homma Y, Li D, Honda F, Aoki D (2021) Inhomogeneous superconducting state probed by 125Te NMR on UTe2. J Phys Soc Jpn 90:064709
16. Sakai H, Tokiwa Y, Opletal P, Kimata M, Awaji S, Sasaki T, Aoki D, Kambe S, Tokunaga Y, Haga Y (2023) Field induced multiple superconducting phases in UTe2 along hard magnetic axis. Phys Rev Lett 130:196002
17. Ran S, Eckberg C, Ding QP, Furukawa Y, Metz T, Saha SR, Liu IL, Zic M, Kim H, Paglione J, Butch NP (2019) Nearly ferromagnetic spin-triplet superconductivity. Science 365:684–687
18. Knafo W, Nardone M, Vališka M, Zitouni A, Lapertot G, Aoki D, Knebel G, Braithwaite D (2021) Comparison of two superconducting phases induced by a magnetic field in UTe2. Commun Phys 4:40
19. Levy F, Sheikin I, Grenier B, Huxley AD (2005) Magnetic field-induced superconductivity in the Ferromagnet URhGe. Science 309:1343
20. Aoki D, Matsuda TD, Taufour V, Hassinger E, Knebel G, Flouquet J (2009) Extremely large and anisotropic upper critical field and ferromagnetic instability in UCoGe. J Phys Soc Jpn 78:113709
21. Tokunaga Y, Aoki D, Mayaffre H, Krämer S, Julien M-H, Berthier C, Horvati x̌ M, Sakai H, Kambe S, Araki S (2015) Reentrant superconductivity driven by quantum tricritical fluctuations in URhGe: evidence from 59Co NMR in URh0.9Co0.1Ge. Phys Rev Lett 114:216401
22. Hattori T, Ihara Y, Nakai Y, Ishida K, Tada Y, Fujimoto S, Kawakami N, Osaki E, Deguchi K, Sato NK, Satoh I (2012) Superconductivity Induced by Longitu dinal Ferromagnetic Fluctuations in UCoGe. Phys Rev Lett 108:066403
23. Wu B, Bastien G, Taupin M, Paulsen C, Howald L, Aoki D, Brison J-P (2017) Pairing mechanism in the ferromagnetic superconductor UCoGe. Nature Commu nications 8:14480
24. Ishida K, Matsuzaki S, Manago M, Hattori T, Kitagawa S, Hirata M, Sasaki T, Aoki D (2021) Pairing interaction in superconducting UCoGe tunable by magnetic field. Phys Rev B 104:144505
25. Duan C, Sasmal K, Maple MB, Podlesnyak A, Zhu JX, Si Q, Dai P (2020) Incommensurate Spin Fluctuations in the Spin-Triplet Superconductor Candidate UTe2. Phys Rev Lett 125:237003
26. Knafo W, Knebel G, Steffens P, Kaneko K, Rosuel A, Brison J, Flouquet J, Aoki D, Lapertot G, Raymond S (2021) Low-dimensional antiferromagnetic fluctuations in the heavy-fermion paramagnetic ladder compound UTe2. Phys Rev B 104:L100409
27. Braithwaite D, Vališka M, Knebel G, Lapertot G, Brison JP, Pourret A, Zhitomirsky ME, Flouquet J, Honda F, Aoki D (2019) Multiple superconducting phases in a nearly ferromagnetic system. Commun Phys 2:147
28. Ran S, Kim H, Liu I-L, Saha SR, Hayes I, Metz T, Eo YS, Paglione J, Butch NP (2020) Enhancement and reentrance of spin triplet superconductivity in UTe2 under pressure. Phys Rev B 101:140503
29. Li D, Nakamura A, Honda F, Sato YJ, Homma Y, Shimizu Y, Ishizuka J, Yanase Y, Knebel G, Flouquet J, Aoki D (2021) Magnetic properties under pressure in novel spin-triplet superconductor UTe2. J Phys Soc Jpn 90:073703
30. Kinjo K, Fujibayashi H, Nakamine G, Kitagawa S, Ishida K, Tokunaga Y, Sakai H, Kambe S, Nakamura A, Shimizu Y, Homma Y, Li D, Honda F, Aoki D (2022) Drastic change in magnetic anisotropy of UTe2 under pressure revealed by 125Te-NMR. Phys Rev B 105
31. Tokunaga Y, Sakai H, Kambe S, Hattori T, Higa N, Nakamine G, Kitagawa S, Ishida K, Nakamura A, Shimizu Y, Homma Y, Li D, Honda F, Aoki D (2019) 125Te NMR study on a single crystal of heavy fermion superconductor UTe2. J Phys Soc Jpn 88:073701

32. Knebel G, Kimata M, Vališka M, Honda F, Li D, Braithwaite D, Lapertot G, Knafo W, Pourret A, Sato YJ, Shimizu Y, Kihara T, Brison J-P, Flouquet J, Aoki D (2020) Anisotropy of the upper critical field in the heavy-Fermion superconductor UTe2 under pressure. J Phys Soc Jpn 89:53707
33. Kohori Y, Kohara T, Shibai H, Oda Y, Kitaoka Y, Asayama K (1988) Nuclear magnetic relaxation in the heavy-Fermion superconductor UPt3. J Phys Soc Jpn 57:395–397
34. Kinjo K, Fujibayashi H, Matsumura H, Hori F, Kitagawa S, Ishida K, Tokunaga Y, Sakai H, Kambe S, Nakamura A, Shimizu Y, Homma Y, Li D, Honda F, Aoki D (2023) Superconducting spin reorientation in spin-triplet multiple superconducting phases of UTe2. Sci Adv 9:eadg2736
35. Imajo S, Kohama Y, Miyake A, Dong C, Tokunaga M, Flouquet J, Kindo K, Aoki D (2019) Thermodynamic investigation of metamagnetism in pulsed high magnetic fields on heavy fermion superconductor UTe2. J Phys Soc Jpn 88:083705
36. Miyake A, Shimizu Y, Sato YJ, Li D, Nakamura A, Homma Y, Honda F, Flouquet J, Tokunaga M, Aoki D (2019) Metamagnetic transition in heavy fermion superconductor UTe2. J Phys Soc Jpn 88:063706
37. Aoki D, Kimata M, Sato YJ, Knebel G, Honda F, Nakamura A, Li D, Homma Y, Shimizu Y, Knafo W, Braithwaite D, Vališka M, Pourret A, Brison J-P, Flouquet J (2021) Field-induced superconductivity near the superconducting critical pressure in UTe2. J Phys Soc Jpn 90:074705

Chapter 6
Summary and Future Work

6.1 Summary

From the ^{125}Te-NMR measurements (Chap. 5), we have investigated the magnetic field and pressure dependence of normal state and superconducting state properties of UTe_2.

6.1.1 Field Induced Superconductivity in UTe_2: Sect. 5.2

We found from the results of *in-situ* χ_{AC} and NMR measurements at magnetic field up to 24.8 T, that the HHSC state is bulk properties of UTe_2, and that the spin component of the triplet pair is along the b axis in the HHSC state, which is different from that in the LHSC state. The results presented here provide decisive evidence that the spin degrees in a spin-triplet pair can be controlled by an external magnetic field H. This is a unique phenomenon that is not expected in spin-singlet superconductivity but is inherent to spin-triplet superconductors. Search of unique phenomena related to the spin degrees of freedom in spin-triplet superconductors is important because this information can facilitate their application to quantum computing.

6.1.2 Superconductivity Emerging in Heavy Fermion State: Sect. 5.3

From the pressure (P) dependence of the NMR results in the $H \parallel b$- and c-axes, it was found that K_b and $1/T_1T$ at low T discontinuously increase at the critical pressure P_c, suggesting the abrupt change in the electronic state and the breakdown of

K. Kinjo, *Nuclear Magnetic Resonance Study on Multiple Superconducting Phases in UTe2*, Springer Theses, https://doi.org/10.1007/978-981-97-8646-6_6

the Kondo coherent state above P_c. We observed the similarity between the pressure and $H \parallel b$ phase diagrams. It was revealed that superconductivity can be realized only in the Kondo coherent state, indicating that the itinerant 5-f nature is crucial for superconductivity in UTe_2. Despite the similarity of the electronic properties in the normal state and the SC properties between UTe_2 and other U-based FM superconductors, the relationship between the magnetic phase and the superconductivity is quite different. Our study clarifies the richness of the spin-triplet superconductivity even in the U-based superconductors.

6.1.3 Superconducting-Spin Rotation: Sect. 5.4

We have performed ^{125}Te-NMR measurement on UTe_2 under pressure to investigate the SC properties and determine the SC order parameter of the SC2 and SC3 phases. The b-axis spin susceptibility in SC2 has the same value as that of the normal state below 3.0 K above 0.5 K. Below 0.5 K, spin susceptibility decreases, and NMR linewidth increases. These results indicate that the SC order parameter is changed, i.e. $\boldsymbol{d}$-vector rotation occurs at the SC2-SC3 transition and the novel SC state in SC3, where spin susceptibility becomes inhomogeneous. This is microscopic evidence for the remaining the spin-degrees of freedom in the SC state of UTe_2, inherent to the spin-triplet superconductivity.

6.1.4 Possible SC Pairing Symmetry

Finally, we discuss the possible SC pairing symmetry by taking into account our recent NMR results obtained in a high-quality single-crystal sample with $T_c = 2.1$ K sample [1]. The data obtained in the 2.1 K sample showed a decrease in the NMR Knight shift in all three crystalline axes (a, b, and c). Considering that the Knight shift decreases when the $\boldsymbol{d}$-vector has a component of the magnetic field direction, the superconducting state is restricted to the A_u state (Table 6.1).

Table 6.1 Classification of the odd-parity SC order parameters for point groups with D_{2h} in a zero field

D_{2h} (zero field)		
IR	Basis functions	SC spin comp.
A_u	$k_a\hat{a}$, $k_b\hat{b}$, $k_c\hat{c}$	
B_{1u}	$k_b\hat{a}$, $k_a\hat{b}$	c
B_{2u}	$k_a\hat{c}$, $k_c\hat{a}$	b
B_{3u}	$k_c\hat{b}$, $k_b\hat{c}$	a

Table 6.2 Classification of odd-parity SC phases occurring in UTe_2 under a b-axis magnetic field

IR of C_{2h} (under field)		
H direction	$A_u^{H\parallel b}$	$B_u^{H\parallel b}$
$H \parallel b$	$A_u + iB_{2u}$	$B_{3u} + iB_{1u}$

First, we consider the high-field superconducting phase. In that case, we need to consider the representation in a magnetic field (Table 6.2). The zero-field superconducting state A_u becomes $A_u^{H\parallel b}$ under the b-axis magnetic field. Given that a field-induced SC-SC transition occurs and that it is 2nd order-like phase transition, the SC phase at high magnetic fields is $B_u^{H\parallel b}$ or spin-singlet superconductivity. However, in these SC states, the spin susceptibility should decrease. This is contrary to our experimental results and physical intuition at a glance. However, Rosuel et al. have proposed that the magnetic field-induced superconducting phase is constantly enhanced by the magnetic field, resulting in the stabilization of singlet superconductivity [2] Their proposal is now controversial, and is checked by the investigation of the normal-state properties above 15 T.

Next, we consider the superconductivity under pressure. The b-axis Knight shift does not decrease in the SC2 state under pressure. This suggests that the SC2 state is the B_{2u} state. This is consistent with the experimental fact that the specific heat jumps twice.

Finally, consider the similarity between pressure effects and magnetic-field effects. If we consider that superconductivity under pressure is a B_{2u} state, this is inconsistent with the similarity to the magnetic field effect. This is because A_u and B_{2u} have the same representation under a b-axis magnetic field and the clear transition experimentally observed is not allowed in principle. To resolve this, the transition between LFSC and HFSC must be a first-order transition or crossover. So far, no such experimental results have been reported. In the 2.5 T data under the pressure, the transition seems to become broader slightly, and this indicates that the $B_{2u} \rightarrow A_u$ transition has become crossover in the b-axis magnetic field, but this is just my speculation.

To resolve these matters, we propose the following future works.

6.2 Future Works

Interesting properties of UTe_2 that were not mentioned in this study are as follows.

1. Investigation of the electronic properties in the normal state above 15 T to elucidate the mechanism that causes superconductivity above 15 T.
2. Properties of SC3 to be elucidated in the lower pressure experiments, especially the temperature dependence of quasiparticle excitations from the measurements of $1/T_1$.

3. NMR measurements to investigate the magnetic fluctuations and superconductivity under pressure for $H \parallel a$.
4. Superconductivity above 40 T.

References

1. in preparation
2. Rosuel A, Marcenat C, Knebel G, Klein T, Pourret A, Marquardt N, Niu Q, Rousseau S, Demuer A, Seyfarth G, Lapertot G, Aoki D, Braithwaite D, Flouquet J, Brison J-P (2022) Thermodynamic evidence for two superconducting phases under magnetic field in UTe2. arXiv:2205.04524

The manufacturer's authorised representative in the EU is Springer Nature Customer Service Centre GmbH, Europaplatz 3, 69115 Heidelberg, Germany. If you have any concerns regarding our products, please contact ProductSafety@springernature.com

Printed and bound by CPI Group (UK) Ltd, Croydon, CR0 4YY
01/10/2026
02225919-0003